UNCERTAINTY ANALYSIS
IN RAINFALL-RUNOFF MODELLING:
APPLICATION OF MACHINE LEARNING TECHNIQUES

T0300020

Uncertainty Analysis in Rainfall-Runoff Modelling:

Application of Machine Learning Techniques

DISSERTATION

Submitted in fulfilment of the requirements of
the Board for Doctorates of Delft University of Technology and
of the Academic Board of UNESCO-IHE Institute for Water
Education for the Degree of DOCTOR
to be defended in public
on Monday, September 28, 2009, at 12:30 hours
in Delft, The Netherlands

by

Durga Lal SHRESTHA
born in Kathmandu, Nepal

Master of Science in Hydroinformatics with Distinction
UNESCO-IHE, the Netherlands

This dissertation has been approved by the supervisors
Prof. dr. D.P. Solomatine
Prof. dr. R.K. Price

Members of the Awarding Committee:

Chairman	Rector Magnificus TU Delft, the Netherlands
Prof. dr. A. Mynett	Vice-Chairman, UNESCO-IHE, the Netherlands
Prof. dr. D.P. Solomatine	TU Delft / UNESCO-IHE, the Netherlands, Supervisor
Prof. dr. R.K. Price	TU Delft / UNESCO-IHE, the Netherlands, Supervisor
Prof. dr. S. Uhlenbrook	VU Amsterdam / UNESCO-IHE, the Netherlands
Prof. dr. N. van de Giesen	TU Delft, the Netherlands
Prof. dr. A. Montanari	University of Bologna, Italy
Prof. dr. J. Hall	Newcastle University, UK
Prof. dr. H.H.G. Savenije	TU Delft, the Netherlands (reserve)

CRC Press/Balkema is an imprint of the Taylor & Francis Group, an informa business

Published by:
CRC Press/Balkema
PO Box 447, 2300 AK Leiden, The Netherlands
e-mail: Pub.NL@taylorandfrancis.com
www.crcpress.com - www.taylorandfrancis.co.uk - www.ba.balkema.nl

ISBN 978-0-415-56598-1 (Taylor & Francis Group)

To
My mother, wife Srijana and son Sujan

SUMMARY

Rainfall-runoff models are widely used in hydrology for a large range of applications and play an important role in optimal planning and management of water resources in river basins. A rainfall-runoff model is, by definition, a simplification of a complex, non-linear, time and space varying hydrological process reality. Such models contain parameters that cannot often be measured directly, but can only be estimated by calibration against a historical record of measured output data. The system input data and output are often contaminated by measurement errors. Consequently predictions made by such a model are far from being perfect and inherently uncertain. It is vital, therefore, that uncertainty should be recognized and properly accounted for. Once the existence of uncertainty in a rainfall-runoff model is acknowledged, it should be managed by a proper uncertainty analysis and by prediction procedures aimed at reducing its impact. There are a number of such procedures actively used but our analysis makes it possible to conclude that they are often based on strong assumptions and suffer from certain deficiencies. This thesis is devoted to developing new procedures for uncertainty analysis and prediction, and testing them on various case studies.

This thesis investigates a number of methods proposed in the literature to provide meaningful uncertainty bounds to the model predictions. Most of the existing methods analyse the uncertainty of the uncertain input variables by propagating it through the deterministic model to the outputs, and hence require the assumption of their distributions and error structures. The majority of the uncertainty methods account for only a single source of uncertainty and ignore other sources of uncertainty explicitly. No single method of uncertainty estimation can be claimed as being perfect in representing uncertainty. Our analysis concludes that the machine learning methods which are able to build accurate models based on data (in this case, on the data about the past model errors) have excellent potential for their use as uncertainty predictors.

Machine learning is concerned with the design and development of algorithms that allow computers to improve their performance over time based on data. A major focus of machine learning techniques is to produce models automatically from data through experience. Over the last 15 years many machine learning techniques have been used to build data-driven rainfall-runoff models. So far these techniques have not been used to provide the uncertainty in the model prediction explicitly in the form of prediction bounds or probability distribution function, particularly in rainfall-runoff modelling. The aim of this research is to explore machine learning techniques to analyse, model and predict uncertainty in rainfall-runoff modelling. We develop two methods, namely the MLUE for parametric and the UNEEC method for residual uncertainty analysis of rainfall-runoff models.

Monte Carlo (MC) simulation is a widely used method for uncertainty analysis in rainfall-runoff modeling and allows the quantification of the model output uncertainty resulting from uncertain model parameters. The MC based methods for uncertainty analysis are flexible, robust, conceptually simple and straightforward; however methods of this type require a large number of samples (or model runs), and their applicability is sometimes limited to simple models. In the case of computationally intensive models, the time and resources required by these methods could be prohibitively expensive. A number of methods have been developed to improve the efficiency of MC based uncertainty analysis methods and still these methods require considerable number of model runs in both offline and operational mode to produce a reliable and meaningful uncertainty estimation. In this thesis we develop a method to predict parametric uncertainty of rainfall-runoff model by building machine learning models that emulate the MC uncertainty results. The proposed method is referred to as the MLUE (Machine Learning in parameter Uncertainty Estimation). The motivation to develop MLUE method is to perform fast parametric uncertainty analysis and prediction.

The MLUE method is applied to a lumped conceptual rainfall-runoff model for the Brue catchment, UK. The generalised likelihood uncertainty estimation method (GLUE) has been used to analyse the parameter uncertainty of the model. We apply the MLUE method to estimate the uncertainty results generated by the GLUE method. We have shown how domain knowledge and analytical techniques are used to select the input data for the machine learning models used in the MLUE method. Three machine learning models, namely artificial neural networks, model trees, and locally weighted regression, are used to predict the uncertainty of the model predictions. The performance of machine learning models in the MLUE method is measured by their predictive capability (e.g., coefficient of correlation, and root mean squared error) and the statistics of the uncertainty (e.g., the prediction intervals coverage probability, and the mean prediction intervals). It is demonstrated that machine learning methods can predict the uncertainty results with reasonable accuracy. The great advantage of the MLUE method is its efficiency to reproduce the MC simulation results; it can thus be an effective tool to assess the uncertainty of flood forecasting in real time.

Generally the analysis of uncertainty consists of propagating the uncertainty of the input and parameters (which is measured by distribution function) through the model by running it for a sufficient number of times and deriving the distribution function of the model outputs. As a part of this research, we develop a novel methodology to analyse, model, and predict the uncertainty of the optimal model output by analysing historical model residuals errors. This method is referred to as UNcertainty Estimation based on Local Errors and Clustering (UNEEC). The UNEEC method consists of the three main steps: (i) clustering the input data in order to identify the homogenous regions of the input space, (ii) estimating the probability distribution of the model residuals for the regions identified by clustering, and (iii) building the machine learning models of the probability distributions of the model errors. Fuzzy clustering has been used to cluster the input data. Three machine learning models, namely artificial neural networks, model trees, and locally weighted regression, are used to predict the uncertainty of the model predictions.

The UNEEC method is applied to rainfall-runoff models of three contrasting catchments: (i) data-driven models for the Sieve catchment, Italy; (ii) lumped conceptual rainfall-runoff model for the Brue catchment, UK; and (iii) lumped conceptual rainfall-runoff model for the Bagmati catchment, Nepal. It has been demonstrated that uncertainty bounds estimated by the UNEEC method are consistent with the order of the magnitude of the model errors. The results show that the percentage of the observed discharges falling within the estimated prediction or uncertainty bounds is very close to the specified confidence level used to produce these bounds with the reasonable width of the bounds. We compare the uncertainty of the model predictions with other uncertainty estimation methods, namely GLUE and two statistical methods - meta-Gaussian, and quantile regression. The comparison results show that the UNEEC method generates consistently narrower uncertainty bounds.

In this research, we have also applied multiobjective optimisation routine NSGA-II to calibrate the HBV rainfall-runoff model for the Bagmati catchment. Four objective functions that emphasise different aspects of the runoff hydrographs are the volume error, root mean squared error (RMSE), RMSE of low flows, and RMSE of peak flows. We implement Pareto preference ordering to select a small number of solutions from the four dimensional Pareto-optimal solutions. The preferred Pareto-optimal set is compared with the optimal values of the parameter set for each of the single objective functions. It is observed that the results of the preferred Pareto-optimal set is a good compromise between all four objective functions and are within the best and worst solutions from the single objective function optimisations. Multiobjective calibration also allows the quantification of the uncertainty in a form of the range of the model simulations corresponding to the Pareto-optimal sets of the parameter; and it provides additional information necessary for risk based decision making.

This research has demonstrated that machine learning methods are able to build reasonably accurate and efficient models for predicting uncertainty, particularly in rainfall-runoff modelling. We conclude that the machine learning techniques can be a valuable tool for uncertainty analysis of various predictive models. The research also presents recommendations for future research and development for uncertainty analysis in the field of rainfall-runoff modelling.

Durga Lal Shrestha

Delft, The Netherlands

Table of Contents

Chapter 1
Introduction

This chapter introduces the subject of this research - uncertainty analysis in rainfall-runoff modelling by applying machine learning. It starts with the background of rainfall-runoff modelling and the treatment of uncertainty analysis within the context of rainfall-runoff modelling. A brief review of uncertainty analysis methods and their shortcomings is also presented. Finally the objectives and the structure of the thesis are presented.

1.1 Background

Rainfall-runoff models are increasingly used in hydrology for a wide range of applications, for example, to extend streamflow records, in the design and operation of hydraulic structures, for real time flood forecasting, to estimate flows of ungauged catchments, and to predict the effect of land-use and climate change. Such models play an important role in water resource planning and management of river basins. These models attempt to simulate complex hydrological processes that lead to the transformation of rainfall into runoff, with varying degrees of abstraction.

A plethora of rainfall-runoff models, varying in nature, complexity, and purpose, has been developed and used by researchers and practitioners in the last century. These rainfall-runoff models encompass a broad spectrum of more or less plausible descriptions of rainfall-runoff relations and processes ranging from the primitive empirical black box model such as the Sherman unit hydrograph method (see, e.g. Sherman, 1932) to the lumped conceptual models such as the Stanford (Crawford and Linsley, 1966), Sacramento (Burnash et al., 1973), HBV (Bergström and Forsman, 1973) models, and the physically based distributed models such as the Mike-SHE model (Abbott et al., 1986a, b). Rapid growth in computational power, the increased availability of distributed hydrological observations and an improved understanding of the physics and dynamics of water systems permit more complex and sophisticated models to be built. While these advances in principle lead to more accurate (less uncertain) models, at the same time, if such complex models with many parameters and data inputs are not parameterized properly or lack input data of reasonable quality, they could be an inaccurate representation of reality.

Since by definition, a rainfall-runoff model is only an abstraction of a complex, non-linear, time and space varying hydrological process of reality, there are many simplifications and idealisations. These models contain parameters that cannot often be measured directly, but can only be estimated by calibration with a historical record of measured output data. The system input (forcing) data such as rainfall, temperature, etc. and output are often contaminated by measurement errors. This inevitably leads to uncertain parameter estimates. Consequently predictions made by such rainfall-runoff model are far from being perfect, in other words, there always exists a discrepancy between the model prediction and the corresponding observed data, no matter how precise the model is and how perfectly the model is calibrated. Thus the model errors which are the mismatch between the observed and the simulated system behaviour are unavoidable in rainfall-runoff modeling due to the inherent uncertainties in the process. Various sources of uncertainty in rainfall-runoff modeling are presented in sections 2.4 and 2.5.

In many fields uncertainty is well recognized and accounted for properly. For example in meteorological sciences, the deterministic weather forecasts or predictions are typically given together with the associated uncertainty. Uncertainty has been also treated in the assessment of the Intergovernmental Panel on Climate Change, IPCC (Swart et al., 2009). In engineering design such as coastal and river flood defenses uncertainty is treated implicitly through conservative design rules, or explicitly by a probabilistic characterization of meteorological events leading to extreme floods. Historically the problem of accurately determining river flows from rainfall, evaporation and other factors was a major focus in hydrology. During the last two decades, there has been a great deal of research into the development and application of (auto) calibration methods (see, e.g., Duan et al., 1992; Solomatine et al., 1999) to improve the deterministic model predictions. Almost all existing river flow simulation techniques are conceived to provide a single estimate, since most research in operational hydrology has been dedicated to finding the best estimate rather than quantifying the uncertainty of model predictions (Singh and Woolhiser, 2002).

It is now being broadly recognized that proper consideration of uncertainty in hydrologic predictions is essential for purposes of both research and operational modeling (Wagener and Gupta, 2005). Along with the recognition of the uncertainty of physical processes, the uncertainty analysis of rainfall-runoff models has become a popular research topic over the past two decades. The value of a hydrologic prediction to water resources and other relevant decision-making processes is limited if reasonable estimates of the corresponding predictive uncertainty are not provided (Georgakakos et al., 2004). Explicit recognition of uncertainty is not enough; in order to have this notion adopted by decision makers in water resources management, uncertainty should be properly estimated and communicated (Pappenberger and Beven, 2006). The research community, however, has done quite a lot in moving towards the recognition of the necessity of complementing point forecasts of decision variables by the uncertainty estimates. Hence, there is a widening recognition of the necessity to (i) understand and identify of the sources of uncertainty; (ii) quantify uncertainty; (iii) evaluate the propagation of uncertainty through the models; and (iv) find means to reduce uncertainty. Incorporating uncertainty into deterministic predictions or forecasts helps to enhance the reliability and credibility of the model outputs.

This dissertation is devoted to developing new methods to analyse model uncertainty, which are based on the methods of machine learning. This study is, in general, in the field of Hydroinformatics, the area that aims in particular at introducing methods of machine learning and computational intelligence into the practice of modelling and forecasting (Abbott, 1991). This study is at the interface between different scientific disciplines: hydrological modelling, statistical and machine learning, and uncertainty analysis.

1.2   Uncertainty analysis in rainfall-runoff modelling

One may observe a significant proliferation of uncertainty analysis methods published in the academic literature, trying to provide meaningful uncertainty bounds of the model predictions. Pappenberger et al. (2006) provide a decision tree to find the appropriate method for a given situation. However, methods to estimate and propagate this uncertainty have so far been limited in their ability to distinguish between different sources of uncertainty and in the use of the retrieved information to improve the model structure analysed. In general, these methods can be broadly classified into six categories (see, e.g., Montanari, 2007; Shrestha and Solomatine, 2008):

1. Analytical methods (see, e.g., Tung, 1996);

2. Approximation methods, e.g., first-order second moment method (Melching, 1992);

3. Simulation and sampling-based (Monte Carlo) methods (see, e.g., Kuczera and Parent, 1998);

4. Methods from group (3) which are also generally attributed to Bayesian methods, e.g., "generalised likelihood uncertainty estimation" (GLUE) by Beven and Binley (1992);

5. Methods based on the analysis of model errors (see, e.g., Montanari and Brath, 2004); and

6. Methods based on fuzzy set theory (see, e.g., Maskey et al., 2004).

Detailed descriptions of these methods are given in section 2.8.

Most of the existing methods (e.g., categories (3) and (4)) analyse the uncertainty of the uncertain input variables by propagating it through the deterministic model to the outputs, and hence require the assumption of their distributions and error structures. Most of the approaches based on the analysis of the model errors require certain assumptions regarding the residuals (e.g., normality and homoscedasticity). Obviously, the relevance and accuracy of such approaches depend on the validity of these assumptions. The fuzzy theory-based approach requires knowledge of the membership function of the quantity subject to the uncertainty which could be very subjective. Furthermore, in majority of the methods, uncertainty of the model output is mainly attributed to uncertainty in the model parameters. For instance, Monte Carlo (MC) based methods analyse the propagation of uncertainty of parameters (measured by the probability density function, pdf) to the pdf of the output. Similar types of analysis are performed for the input or structural uncertainty independently. Methods based on the analysis of the model errors typically compute the uncertainty of the "optimal model" (i.e. the model with the calibrated parameters and the fixed structure), and not of the "class of models" (i.e. a group of models with the same structure but parameterised differently) as, for example, MC methods do.

The contribution of various sources of errors to the total model error is typically not known and, as pointed out by Gupta et al. (2005), disaggregation of errors into their source components is often difficult, particularly in hydrology where models are non-linear and different sources of errors may interact to produce the measured deviation. Nevertheless, evaluating the contribution of different sources of uncertainty to the overall uncertainties in model prediction is important, for instance, for understanding where the greatest sources of uncertainties reside, and, therefore directing efforts towards these sources (Brown and Heuvelink, 2005). In general, relatively few studies have been conducted to investigate the interaction between different sources of uncertainty and their contributions to the total model uncertainty (Engeland et al., 2005; Gupta et al., 2005). For the risk based decision-making process such as flood warnings, it is more important to know the total model uncertainty accounting for all sources of uncertainty than the uncertainty resulting from individual sources.

However, the practice of uncertainty analysis and use of the results of such analysis in decision making is not widespread, for several reasons (Pappenberger and Beven, 2006). Uncertainty analysis takes time, so adds to the cost of risk analysis, options appraisal and design studies. It is not always clear how uncertainty analysis will contribute to improved decision making. Much of the academic literature on hydrological uncertainties (Liu and Gupta, 2007) has tended to focus upon forecasting problems. Identifying uncertainty bounds on a flood forecast is important, but to be meaningful needs to be set within the context of a well defined decision problem (Frieser et al., 2005; Todini, 2008). The uncertainty analysis requires careful interpretation in order to understand the meaning and significance of the results. It is through this process of scrutiny and discussion that the most useful insights for decision makers are obtained. Furthermore, the conduct of uncertainty analysis provides new insights into model behaviour that will need to be discussed and agreed with the experts responsible for models that are input into the analysis. Indeed the process of scrutiny that uncertainty analysis provides is an additional benefit (Hall and Solomatine, 2008).

Experience is now growing in the communication of uncertainty to decision makers and members of the public, for example in the context of environmental risks (Sluijs et al., 2003) and climate change (IPCC, 2005). Sluijs et al. (2003) stress the importance of engaging stakeholders from this early stage, identifying the target audiences and then using appropriate language to communicate uncertainties. Alongside numerical results and their implications for decision makers, the limitations of data sources and analysis methods should be made clear and areas of ignorance should be highlighted.

1.3   Machine learning in uncertainty analysis

Over the last 15 years many machine learning techniques have been used extensively in the field of rainfall-runoff modelling (see section 3.4 for more detail). These techniques have been also used to improve the accuracy of prediction/forecasting made by process based rainfall-runoff models. Generally, they are used to update the output variables by forecasting the error of the process based models. All these techniques to update the model predictions can be seen as error modelling paradigm to reduce the uncertainty of the predictions. However, these techniques do not provide explicitly the uncertainty of the model prediction in the form of prediction bounds or probability distribution function of the model output.

In this thesis we explore the possibility of using machine learning techniques which can provide the reasonable uncertainty estimation for the runoff prediction made by rainfall-runoff models.

As mentioned above in section 1.1, with advances in computational power and technological development, more complex and sophisticated distributed rainfall-runoff models have been built and used in practice. Computational burden in distributed runoff models is now less problematic than before, although it still can be an issue when predictive uncertainty of a model is assessed through laborious MC simulations (Beven, 2001). Several uncertainty analysis methods based on MC simulations have been developed to propagate the uncertainty through the models. The MC based method for uncertainty analysis of the outputs of such models is straightforward, but becomes impractical in real time applications for computationally intensive complex models when there is insufficient time to perform the uncertainty analysis because the large number of model runs is required. Practical implementation of MC based uncertainty analysis methods face two major problems: (i) convergence of the MC simulations is very slow (with the order of computational complexity (O) of $1/\sqrt{s}$), so a large number of runs needed to establish a reliable estimate of uncertainties; and (ii) the number of simulations increases exponentially with the dimension of the parameter vector ($O(n^p)$) to cover the entire parameter domain, where s is the number of simulations, p is the dimension of parameter vector, n is the number of samples required for each parameter.

A number of research have been conducted to improve the efficiency of MC based uncertainty analysis methods such as Latin hypercube sampling (McKay et al., 1979), and the moment propagation techniques (Rosenblueth, 1975; Harr, 1989; Melching, 1992). However all these methods require running the model many times in both offline and online mode. In other words, MC based methods require running the models in a loop each time when the uncertainty of the model prediction for the new input data $x_{(T+1)}$ is required. In this thesis we explore an efficient method to assess the uncertainty of the model M for $t = T+1$ when new input data $x_{(T+1)}$ is feed. The method we propose encapsulates the MC based uncertainty results in machine learning models and is referred to as a "**M**achine **L**earning in parameter **U**ncertainty **E**stimation" (MLUE). In the MLUE method, the machine learning model is used as a surrogate model to emulate the laborious MC based uncertainty methods and hence provides an approximate solution to the uncertainty analysis in a real time application without re-running the MC runs. Surrogate modelling is the process of constructing approximation models (emulators) that mimic the behavior of the simulation model as closely as possible while being computationally cheap(er) to run. We believe that it is preferable to have an approximate uncertainty estimate than no uncertainty estimate at all.

Yet another problem relates to the situation when an interest is in assessing model uncertainty when it is difficult to attribute it to any particular source. If the data and resources are available and the computational time allows to do a full MC based uncertainty analysis method, then it is preferable to perform the latter. However in practice, engineering decisions are often based on a single (optimal) model run without any uncertainty analysis. In this thesis we also develop a novel method for uncertainty analysis of a calibrated model based on the historical model residuals. The historical model residuals (errors) between the model prediction and the observed data are the best available quantitative indicators of the discrepancy between the model and the real-world system or process, and they provide valuable information that can be used to assess the predictive uncertainty. The residuals and their distribution are often functions of the model input variables and can be predicted by building a separate model mapping of the input space to the model residuals or even their probability distribution function. In other words, the idea here is to learn the relationship between the probability distribution of the model residuals and the input variables; and to use this relationship to predict the uncertainty of the model prediction of the output variable (e.g., runoff) in the future. This approach is referred to as an ''**UN**certainty **E**stimation based on local **E**rrors and **C**lustering'' (UNEEC). The UNEEC method estimates the uncertainty of the optimal model that takes into account all sources of errors without attempting to disaggregate the contribution given by their individual sources. The UNEEC method is based on the concept of optimality instead of equifinality as it analyzes the historical model residuals resulting from the optimal model (both in structure and parameters).

1.4   Objective of this study

The aim of this research is to develop methodology for uncertainty analysis in rainfall-runoff modelling using machine learning techniques. The objectives of the research are:

1. To review the existing methods of uncertainty analysis in rainfall-runoff modelling;

2. To review machine learning methods and to investigate the possibility of applying machine learning methods in uncertainty analysis;

3. To develop a methodology for uncertainty analysis in rainfall-runoff modeling using machine learning methods;

4. To develop a methodology for the surrogate modeling of uncertainty generated by the Monte Carlo based uncertainty methods; and

5. To implement the developed methodologies in computer codes and to test the methodologies by application to real-world problems.

1.5 Outline of the thesis

The thesis is organised in eight chapters. A brief overview of the structure is given below. Chapter 2 is devoted to a review of uncertainty analysis especially in rainfall-runoff modelling. It starts with brief overviews of rainfall-runoff models and their classification which is followed by the sources of uncertainty in the context of rainfall-runoff modelling. It also discusses the commonly used uncertainty representation theories of probability, fuzzy logic and entropy. It briefly reviews the various uncertainty analysis methods used in rainfall-runoff modelling.

Chapter 3 presents several machine learning techniques used in this study. It describes artificial neural networks, model trees, instance based learning and clustering techniques.

In Chapter 4, a novel method "Machine Learning in parameter Uncertainty Estimation" (MLUE) to modelling parametric uncertainty of rainfall-runoff models is presented. It is observed that there exists a dependency between the forcing input data, the state variables of rainfall-runoff models and the uncertainty of the model predictions. Chapter 4 explores building machine learning models to approximate the functional relationship between the input data (including state variables if any) and the uncertainty of the model prediction such as a quantile.

Chapter 5 presents the application of the MLUE method for parametric uncertainty representation and analysis. Various machine learning models such as artificial neural networks, model trees, locally weighted regression, have been used. The MLUE method is applied to analyse the uncertainty of a lumped conceptual rainfall-runoff model of the Brue catchment in the UK.

Chapter 6 presents a novel method "Uncertainty Estimation based on Local Error and Clustering" (UNEEC) for uncertainty analysis of rainfall-runoff models. This method assumes that the model residuals or errors are indicators of the total model uncertainty. The method estimates the "residual uncertainty" of the optimal model that takes into account all sources of errors without attempting to disaggregate the contribution given by their individual sources.

Chapter 7 provides the application of the UNEEC methodology to estimate uncertainty of rainfall-runoff models in a number of catchments. In the first part the UNEEC method is applied to estimate uncertainty of the forecasts made by several machine learning methods in the Sieve catchment in Italy. The second part covers the application to estimate uncertainty of the conceptual rainfall-runoff models of two catchments: the Brue in UK and the Bagmati in Nepal. The comparison results with other uncertainty methods are presented as well.

Chapter 8 presents the conclusions of the research based on the various case studies presented in this thesis. Finally the possible directions for further research are suggested.

Chapter 2
Uncertainty Analysis in Rainfall-Runoff Modelling

This chapter presents the three classes of rainfall-runoff models: data-driven, conceptual and physically-based. Classification and sources of uncertainty in the context of rainfall-runoff modelling are described. Sources of uncertainty in data-driven modelling are also presented. The chapter also discusses the commonly used uncertainty representation theories of probability, fuzzy logic and entropy. Brief reviews of the uncertainty analysis methods used in rainfall-runoff modelling are given.

2.1 Types of rainfall-runoff models

Many attempts have been made to classify rainfall-runoff models, see, e.g., Clarke (1973), Todini (1988), Chow et al. (1988), Singh (1995b), Refsgaard (1996). The classifications are generally based on the following criteria: (i) the extent of physical principles that are applied in the model structure; (ii) the treatment of the model inputs and parameters as a function of space and time. According to the first criterion (i.e., physical process description), a rainfall-runoff model can be attributed to two categories: deterministic and stochastic (see Figure 2.1). A deterministic model does not consider randomness; a given input always produces the same output. A stochastic model has outputs that are at least partially random.

Deterministic models can be classified according to whether the model gives a lumped or distributed description of the considered sub catchment area (i.e., second criterion) and whether the description of the hydrological processes is empirical, conceptual or more physically based (Refsgaard, 1996). On this basis deterministic rainfall-runoff models are classified as: (i) data-driven models (black box), (ii) conceptual models (grey box); and (iii) physically based models (white box). Subsections 2.1.1-2.1.3 present these three types of rainfall-runoff models.

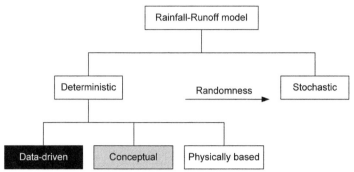

Figure 2.1. Classification of rainfall-runoff models according to physical processes (adapted from Refsgaard, 1996).

2.1.1   Data-driven models

Data-driven (also called metric, or empirical, or black box) models involve mathematical equations that have been derived not from the physical processes in the catchments but from an analysis of the concurrent input and output time series. Typically such models are valid only within the boundaries of the domain where data is given (Price, 2002). Black box models may be divided into three main groups according to their origin, namely empirical hydrological methods, statistically based methods and newly developed Hydroinformatics based methods.

Unit hydrograph methods originally introduced by Sherman (1932) are widely used examples of empirical methods. Although the derivation of these types of methods is purely empirical, the selection of the input-output parameters is by and large, dictated by some physical understanding of the processes.

Statistically based methods consist of regression and correlation models. Linear regression and correlation techniques are standard statistical methods used to determine the functional relationship between input and output. The autoregressive integrated moving average (ARIMA) model (Box and Jenkins, 1970) is an extensively used method in rainfall-runoff modelling and is often attributed to stochastic model.

Hydroinformatics based methods also simply called data-driven models are relatively new methods which are generally based on machine learning techniques (see Chapter 3). Examples of such methods are artificial neural networks, fuzzy regression (Bardossy et al., 1990), model trees (Quinlan, 1992), genetic programming, support vector machines (Vapnik, 1995, 1998), instance based learning. Over the past ten years these techniques have been used extensively in rainfall-runoff modelling (see, e.g., Minns and Hall, 1996; Abrahart and See, 2000; Govindaraju and Rao, 2000; Solomatine and Dulal, 2003; Solomatine et al., 2008). Data-driven techniques are presented in Chapter 3.

2.1.2   Conceptual models

Conceptual models are generally composed of a number of interconnected storages representing physical elements in a catchment. These storages are recharged through fluxes of rainfall, infiltration or percolation and depleted through evapotranspiration, runoff etc. assembling the real physical process in the catchment. Parameters and fluxes typically represent the average values over the entire catchment. The equations used to describe the process are semi-empirical, but still with a physical basis. The model parameters cannot usually be assessed from field data alone, but have to be obtained through the help of calibration. Although the conceptual models are simple and can be easily implemented in the computer code, they need sufficiently long meteorological and hydrological records for their calibration which may not be always available. The calibration of the conceptual models involves curve fitting, thus making physical interpretation of the fitted parameter very difficult and predicting effects of land use change by changing parameter values cannot therefore be done with confidence (Abbott et al., 1986a).

There are many conceptual models with different levels of physical representational and varying degree of complexity. Crawford and Linsley (1966) are credited for the development of the first major conceptual model by introducing the well known Stanford Watershed Model IV. Numerous other widely used conceptual models include Sacramento Soil Moisture Accounting model (Burnash et al., 1973), NAM model (Nielsen and Hansen, 1973), TOPMODEL (Beven and Kirkby, 1979), TANK model (Sugawara, 1967, 1995), HBV model (Bergström and Forsman, 1973) and so on. A brief description of several conceptual models is given in an early work by Fleming (1975). Comparison results of 10 different conceptual models used in the sixties for operational hydrological forecasting are presented in WMO (1975). More comprehensive descriptions of a large number of conceptual models are provided in Singh (1995a).

2.1.3   Physically based models

Physically based model are based on the general principles of physical processes (e.g. continuity, momentum and/or energy conservation) and describe the system behaviour in as much detail as conceivable. The state and evolution of the system is described using state variables that are functions of both space and time. These variables bear physical meaning and most of them are measurable. In such models, the hydrological processes of water movement are modelled by finite difference representations of the partial differential equations of mass, momentum and energy conservation, or by empirical equations derived from independent experimental research (Abbott et al., 1986b).

Although physically based models do not in principle require lengthy hydrometeorological data for their calibration, they do require the evaluation of a large number of parameters describing the physical characteristics of the catchment on a spatially distributed basis (Abbott et al., 1986a) and huge amounts of data about the initial state of the model (e.g. initial water depth, soil moisture content and depth to the aquifer water table at all points of a catchment) and the description of the morphology (topography, topology and dimensions of the river network, geometry of the geological layers, etc.) of the system to be modeled. Most of the time, however, such data are not available. Furthermore, physically based models suffer from scale related problems (e.g., most measurements being made at the point scale, whereas model parameters are required at the scale of the grid network used in the mathematical representation of the catchment) and over-parameterisation (Beven, 1989). Such models are complex and their development requires considerable resources in human expertise and computing capability. However, physically based models can overcome many of the deficiencies mentioned in section 2.1.2 through their use of parameters which have a physical interpretation and through their representation of spatial variability in the parameter values (Abbott et al., 1986a). The principles used in such models are assumed to be valid for a wide range of situations including those that have not yet been observed (Guinot and Gourbesville, 2003). Physically based models can also generate a large amount of information on what happens away from the boundaries (where data is prescribed as input or output) (Price, 2006). A typical example of a physically based hydrological modelling system is the SHE/ MIKE SHE system (Abbott et al., 1986a, b).

2.1.4 Stochastic models

If any of the input-output variables or error terms of the model are regarded as random variables having probability distribution, then the model is stochastic. This definition is in accord with Chow (1964): "if the chance of occurrence of the variables is taken into consideration and the concept of probability is introduced in formulating the model, the process and the model are described as stochastic or probabilistic."

An example of a stochastic model is Markov model for streamflow of order one as given by (see, Fleming, 1975):

$$Y_i = \mu + \rho(Y_{i-1} - \mu) + R_i \sigma \sqrt{1 - \rho^2}$$

(2.1)

where Y_i is the streamflow in period i, μ is the mean of data, ρ is the lag-one serial correlation, σ is the standard deviation of the streamflow, and R is a independent and identically distributed random variable with zero mean and standard deviation 1. ARIMA (Box and Jenkins, 1970) is another example of stochastic model. Examples of other stochastic models can be found in Clarke (1973).

2.2   Notion of uncertainty

It is difficult to find in literature a precise and generally applicable definition of the term *uncertainty*. It seems that there is no consensus within the profession about the very term of uncertainty, which is conceived with differing degree of generality (Kundzewicz, 1995). The meanings of the word *uncertain* as given by Webster's (1998) dictionary are the following: not surely or certainly known, questionable, not sure or certain in knowledge, doubtful, not definite or determined, vague, liable to vary or change, not steady or constant, varying. The noun "uncertainty" results from the above concepts and can be summarized as the state of being uncertain.

Uncertainty can be defined with respect to *certainty*. For example, Zimmermann (1997) defined certainty as: "certainty implies that a person has quantitatively and qualitatively the appropriate information to describe, prescribe or predict deterministically and numerically a system, its behaviour or other phenomena." Situations that are not described by the above definition shall be called uncertain. Similar definition given by Gouldby and Samuels (2005): "a general concept that reflects our lack of sureness about someone or something, ranging from just short of complete sureness to an almost complete lack of conviction about an outcome".

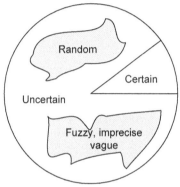

Figure 2.2. Certainty and uncertainty in the information world (source: Ross, 1995).

Ross (1995) illustrated in a diagram (Figure 2.2) that the presence of uncertainty in a typical problem is much larger compared to the presence of certainty. The collection of all information in this context is termed as the *information world* (represented by the circle in Figure 2.2). If the uncertainty is considered in the content of this information, only a small portion of a typical problem might be regarded as certain. The rest inevitably contains uncertainty that rises from the complexity of the system, from ignorance, from chance, from various classes of randomness, from imprecision, from lack of knowledge, from vagueness, and so forth.

Regarding the occurrence of uncertainty, Klir and Wierman (1998) stated:

When dealing with real-world problems, we can rarely avoid uncertainty. At the empirical level, uncertainty is an inseparable companion of almost any measurement, resulting from a combination of inevitable measurement errors and resolution limits of measuring instruments. At the cognitive level, it emerges from the vagueness and ambiguity inherent in natural language. At the social level, uncertainty has even strategic uses and it is often created and maintained by people for different purposes (privacy, secrecy, propriety).

In the context of the present research work, uncertainty is defined as a state that reflects our lack of sureness about the outcome of physical processes or system of interest and gives rise to a potential difference between assessment of the outcome and its "true" value. More precisely, uncertainty of a model output (for example, water level at the bridge) is the state or condition that the output can not be assessed uniquely. Uncertainty stems from incompleteness or imperfect knowledge or information concerning the process or system in addition to the random nature of the occurrence of the events. Uncertainty resulting from insufficient information may be reduced if more information is available.

The following sections attempt to describe uncertainty and distinguish sources of uncertainty and types of uncertainty in general and in the context of rainfall-runoff modelling. Different theories of uncertainty representation and modelling are described in sections 2.7 and 2.8.

2.3   Classification of uncertainty

There were numerous attempts in the literatures to classify and distinguish different types of uncertainty. Yen and Ang (1971) classified uncertainties into two types: *objective* uncertainties associated with any random process or deductible from statistical samples and *subjective* uncertainties for which no quantitative factual information is available. Burges and Lettenmaier (1975) categorized two types of uncertainties: type I error results from the use of an inadequate model with correct parameter values, and type II error assumes the use of a perfect model with parameter subject to uncertainty.

Klir and Folger (1988) classified uncertainty into two types: *ambiguity* and *vagueness*. Ambiguity is the possibility of having multiple outcomes for processes or systems, and vagueness is the non-crispness of belonging and non-belonging of elements to a set of or a notion of interest. This classification is consistent with the classification given by Ayyub and Chao (1998).

Kaufmann and Gupta (1991)classified uncertainty as *random* and *fuzzy*. Following this, Davis and Blockey (1996) identified three distinct components of uncertainty: *fuzziness, incompleteness* and *randomness.* In the context of modelling engineering system, Van Gelder, (2000) and Hall (2003) classified uncertainty into *inherent* and *epistemic*. The *inherent* or *aleatory* uncertainty represents the randomness and variability observed in nature (both in space and time), whereas *epistemic* or *knowledge* uncertainty refers to the state of knowledge of a physical system and our ability to measure and model. Abebe (2004) classified uncertainty into *structured* and *unstructured* based on the characteristics of the gap between the model and the physical system as it is revealed in the form of errors. A comprehensive review of the classification of uncertainty can be found in Hall (1999).

2.4   Sources of uncertainty in rainfall-runoff models

The uncertainties in rainfall-runoff modelling stem mainly from the three important sources (see, e.g., Melching, 1995; Refsgaard and Storm, 1996):

Observational uncertainty: Uncertainty related to the observation used for rainfall-runoff modelling can be classified as observational uncertainty. The observation is the measurement of the input (e.g., precipitation and temperature) and output (e.g., streamflow) fluxes of the hydrological systems and sometimes of its states (e.g., soil moisture content, groundwater levels). The observational or data uncertainty usually consists of two components: a) measurement errors due to both instrumental and human errors b) error due to inadequate representativeness of a data sample due to scale incompatibility or difference in time and space between the variable measured by the instrument and the corresponding model variable. The latter is sometimes called *sampling uncertainty* (see, e.g., Abebe, 2004). These two error components have very different characteristics which may vary from variable to variable; hence, statistics of both errors should be considered and adequately specified.

Model uncertainty: A model is a simplified representation of the real world (Refsgaard, 1996). The complex real processes are greatly simplified while deriving the basic concepts and equations of the model. Conceptualisation with inappropriate approximations and ignored or misrepresented processes can result in large error in the conceptual structure of the model. Model errors can also arise from the mathematical implementation (e.g., spatial and temporal discretisations) that transforms a conceptual model into a numerical model.

Parameter uncertainty: The uncertainty in the model parameters results from an inability to accurately quantify the input parameters of a model. The parameters of the model may not have direct physical meaning. Furthermore, those parameters that have a physical meaning can not be directly measured or it is too costly to measure them in the field. In such case their values are generally estimated by indirect means (e.g., expert judgment, model calibration etc). Expert judgment is normally subjective and hence uncertain. The parameters obtained from the calibration process are also not free from uncertainty for various reasons including data uncertainty (data used to calibrate the model contains errors), model uncertainty (the model structure used to calibrate the model is not adequate), lack of sufficient data etc.

2.5   Sources of uncertainty in context of data-driven modelling

In section 2.4, we discussed the source of uncertainty in physically based rainfall-runoff models. The sources of uncertainty in data-driven rainfall-runoff modeling are presented in this section. The model predictions from data-driven techniques are uncertain due to the following sources:

Inaccuracies in the training data (data uncertainty): The training data set is typically noisy due to systematic and random errors. Furthermore, it is incomplete because all the possible input-output examples are not available. Indeed the training data is only a random sample of an unknown distribution from population. Noise is inherent to all real data and contributes to the total prediction variance as data noise variance σ_v^2.

Limitation of the regression model (model uncertainty): The limitations of the model (i.e., capacity) due to the training algorithm introduce uncertainty to the predictions. For example, fitting the non-linear relationship between variables by linear regression introduces uncertainty in the predictions. Similarly, most of the data-driven techniques are trained using an iterative optimisation algorithm. The resultant coefficients or parameter values of the model often correspond to a local rather than global optimum of the error function. Such types of uncertainties are often called model uncertainties and contribute to the total prediction variance as model uncertainty variance σ_m^2.

Non optimal parameters of the model: This is analogous to the parameter uncertainty in physically based model. There are different parameters (for example learning rate, momentum etc. in neural networks) that have to be selected for optimal results. If these parameters are not optimal, then they may lead to different outcomes.

Not optimal splitting of data: A standard procedure for evaluating the performance of a model would be to split the data into training set, cross-validation set and test set. This approach is however, very sensitive to the specific sample splitting (LeBaron and Weigend, 1994). In principle, all these splitting data sets should have identical distributions, but the true distribution of the data is not known in advance. So this causes uncertainty in prediction as well.

The prediction error of a regression model can be decomposed into the following three sources (Geman et al., 1992): (i) model bias; (ii) model variance; and (iii) noise.

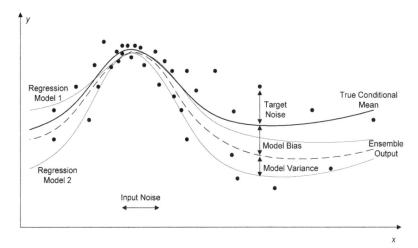

Figure 2.3. Model bias and variance. Dots represent the observed variable.

Model bias and *variance* may be further decomposed into contributions from data and contributions from the training process. Furthermore, noise can be also decomposed into target noise and input noise. The decomposition of the prediction error is illustrated in Figure 2.3. Estimating these components of prediction error (which is however not always possible) helps to compute the predictive uncertainty.

The terms bias and variance come from a well-known decomposition of prediction error. Given n data points and M models, the decomposition is based on the following equality:

$$\frac{1}{nM}\sum_{i=1}^{n}\sum_{j=1}^{M}(y_i - \hat{y}_{ij})^2 = \frac{1}{n}\sum_{i=1}^{n}(y_i - \bar{\hat{y}}_i)^2 + \frac{1}{nM}\sum_{i=1}^{n}\sum_{j=1}^{M}(\bar{\hat{y}}_i - \hat{y}_{ij})^2 \qquad (2.2)$$

where y_i denotes the ith target, $\hat{y}_{i,j}$ denotes the ith output of the jth model, and

$\bar{\hat{y}}_i = \dfrac{1}{M}\sum_{j=1}^{M}\hat{y}_{ij}$ denotes the average model output calculated for input i.

The left hand term of equation (2.2) is the well known mean squared error. The first term in the right hand site is the square of bias and the last term is the variance. The bias of the prediction errors measures the tendency of over or under prediction by a model, and is the difference between the target value and the model output. From equation (2.2) it is clear that the variance does not depend on the target, and measures the variability in the predictions by different models.

2.6 Uncertainty analysis in rainfall-runoff modelling

As discussed in section 2.4 and Chapter 1, it should be recognized that a model is only an abstraction of reality, which generally involves certain degrees of simplifications and idealisations. Hence the model predictions are far from being perfect no matter how sophisticated the models are, and subject to different degree of uncertainty. As mentioned above uncertainty in the model predictions results mainly from structure, input and parameter uncertainty. Figure 2.4 shows how different sources of the uncertainty vary with model complexity. In this context model complexity is measured by its number of parameters and requirements of input data. As the model complexity increases, in principle, structure uncertainty decreases due to detailed representation of the physical process. However, with the increasing complexity of model, the number of the input and parameters also increases; and if such complex models with many parameters and data inputs are not parameterized properly or lack quality input data, there is a high probability that uncertainty associated with input data or parameter estimation will increase. Due to the inherent trade-off between model structure uncertainty and input/parameter uncertainty, for every model and the associated data set there exists the optimal level of model complexity where the total uncertainty is minimum. A key question that then arises is how to find the optimal level of model

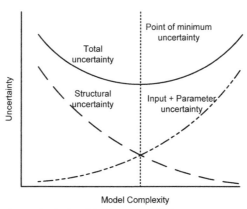

Figure 2.4. An illustration of relationship between different sources of uncertainty and the combined effect on total model uncertainty. Note that y-axis scale does not start at zero.

complexity; nevertheless, several methods have been developed for this purpose (see, e.g., Schoups et al., 2008).

Once the existence of uncertainty in a rainfall-runoff model is acknowledged, it should be managed by a proper uncertainty analysis process to reduce the impact of uncertainty. There is a large number of uncertainty analysis methods published in the academic literature and Pappenberger et al. (2006) provide a decision tree to find the appropriate method for a given situation. The uncertainty analysis process in rainfall-runoff models varies mainly in the following ways: (i) the type of rainfall-runoff models used; (ii) the source of uncertainty to be treated; (iii) the representation of uncertainty; (iv) the purpose of the uncertainty analysis; and (v) the availability of resources. Uncertainty analysis is a well-accepted procedure and has comparatively long history in physically based and conceptual modelling (see, e.g., Beven and Binley, 1992; Gupta et al., 2005). With increasing application of data-driven techniques in rainfall-runoff modelling, there is also growing interest to quantify the uncertainty of the predictions made by the data-driven models. The uncertainty methods used in such models are obviously different from those used in physically based and conceptual models.

Regarding to the sources of uncertainty, MC type methods are widely used for parameter uncertainty, Bayesian methods and/or data assimilation can be used for input uncertainty and Bayesian model averaging method is suitable for structure uncertainty. Uncertainty analysis method also depends on whether the uncertainty is represented as randomness or fuzziness for example (see section 2.7). Similarly, uncertainty analysis methods for real time forecasting purpose would be different than those used in design purpose (design discharge for structures built in the rivers). In the former case the running time of the model is crucial and hence computationally expensive MC based methods are impractical. Availability of resources such as computational power also determines the selection of different methods used for uncertainty analysis.

Uncertainty analysis methods in all of the above cases should involve: (i) identification and quantification of the sources of uncertainty; (ii) reduction of uncertainty; (iii) propagation of uncertainty through the model; (iv) quantification of uncertainty in the model outputs; and (vi) application of the uncertain information in decision making process. However, the practice of uncertainty analysis and use of the results of such analysis in decision making is not widespread. The reasons for this are stated by Pappenberger and Beven (2006):

"(i) uncertainty analysis is not necessary given physically realistic models; (ii) uncertainty analysis is not useful in understanding hydrological and hydraulic processes; (iii) uncertainty (probability) distributions cannot be understood by policy makers and the public; (iv) uncertainty analysis cannot be incorporated into the decision-making process; (v) uncertainty analysis is too subjective; (vi) uncertainty analysis is too difficult to perform; and (vii) uncertainty does not really matter in making the final decision".

For details the readers are referred to the citation above.

2.7   Uncertainty representation

Prior to the entry of *fuzzy set theory*, the *probability theory* has been the primary tool for representing uncertainty in mathematical models. Klir and Folger (1988) suggest that probability theory is appropriate for dealing with only a very special type of uncertainty, namely randomness. However, not all uncertainty is random and probability distribution may be not the best way to describe it. As mentioned in section 2.3, some forms of uncertainty are due to vagueness or imprecision and cannot be treated with probabilistic approaches. Fuzzy set theory and *fuzzy measures* (Zadeh, 1965, 1978) provide a non-probabilistic approach for modelling the kind of uncertainty associated with vagueness and imprecision.

Information theoretic approach is also used for measuring uncertainty. Information is that which resolves and/or reduces uncertainty (Shannon, 1948). *Shannon's entropy* belongs to this type of approach. Entropy is a measure of uncertainty and information formulated in terms of probability theory. Another broad theory of uncertainty representation is the *evidence theory* introduced by Shafer (1976). Evidence theory also known as *Dempster-Shafer theory of evidence* is based on both probability and possibility theory. The following sub sections provide some details of probability theory, fuzzy set theory and entropy.

2.7.1   Probability theory

Of all the methods for uncertainty analysis, *probabilistic methods* have by far the longest tradition and are the best understood. This of course does not imply that they are beyond criticism as any method of handling uncertainty. It does, however, mean that they are relatively well tested and well developed and can act as a standard against which other more recent approaches may be measured (Krause and Clark, 1993). There are basically two broad views of probability concept: *frequentist view* and *subjective view*. These are discussed in the following.

The frequency view of probability relates to the situation where an experiment can be repeated indefinitely under essentially identical conditions, but the observed outcome is random (not the same every time). Empirical evidence suggests that the proportion of times any particular event has occurred, i.e. its *relative frequency*, converges to a limit as the number of repetitions increases. This limit is the probability of the event (Hall, 2003). Thus, if n_t is the total number of experiments and n_x is the number of times where the event x occurred, the probability $P(x)$ of the event occurring will be approximated by the relative frequency as follows:

$$P(x) \approx \frac{n_x}{n_t} \qquad (2.3)$$

Mentioned above, as the number of experiments approaches infinity, the relative frequency will converge exactly to the probability:

$$P(x) = \lim_{n_t \to \infty} \frac{n_x}{n_t} \qquad (2.4)$$

In subjective view, the probability is interpreted as a degree of belief. Subjective view of probability is also referred as *Bayesian probability*, in honour of Thomas Bayes (1702-1761), a scientist from the mid-1700s who helped to pioneer the theory of probabilistic inference. The Bayesian probability of an event x is a person's degree of belief in that event. Whereas a classical probability is a physical property of the world (e.g., the probability that a coin will land heads), the Bayesian probability is a property of the person who assigns the probability (e.g., our degree of belief that the coin will land heads). The basic idea in the application of subjective view of probability is to assign a probability to any event on the basis of the current state of knowledge and to update it in the light of the new information. The conventional procedure for updating a prior probability (i.e. initial belief) in the light of new information is by using the famous Bayes' theorem. It is presented later in this section.

The mathematics of the probability theory is based on three basis axioms formulated in 1933 by Andrei N. Kolmogorov. Let us consider the universal set X containing all possible elements of concern. Let the event A be an element of the universal set X. The probability $P(A)$ of an event A is a real number in the unit interval [0,1]. The probability obeys the following basic axioms:

Axiom 1: $P(A) >= 0$, i.e., the probability of an event is always non-negative.
Axiom 2: $P(X) = 1$, i.e. the probability of the universal set is 1.
Axiom 3: For any sequences of disjoint sets $A_i \in X$
$$P(A_1 \cup A_2 \cup ...) = \sum_i P(A_i)$$

Probability distribution

Let the random variable X has a set of possible discrete states $x_1,...,x_n$. Then $P(X = x_i)$ denotes the probability of variable X being in state x_i and it follows from the axioms that:

$$\sum_{i=1}^{n} P(X = x_i) = 1$$ (2.5)

The sequence of probabilities $P(X = x_1), ..., P(X = x_n)$ defines a probability vector $P(X)$ which represents the probability distribution of X over all of its states. In probability theory, the uncertainty of the random variable X is represented by its *probability distribution* over all of its states. *Cumulative distribution function* (cdf) is the accumulation of the probability distributions defined as:

$$F_X(x) = P(X \le x) = \int_{-\infty}^{x} f_X(x)dx$$ (2.6)

where $f_X(x)$ is the probability density function of the random variable X at x (see Figure 2.5). CDF at a (Figure 2.5 (b)) is given by

$$F_X(b) = P(X \le b) = \int_{-\infty}^{b} f_X(x)dx$$ (2.7)

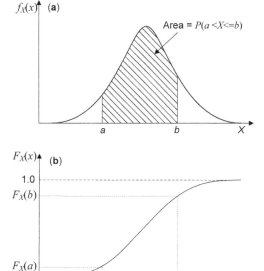

Figure 2.5. Example of (a) probability density function and (b) cumulative distribution function.

It then follows

$$P(a < X \le b) = \int_{-\infty}^{b} f_X(x)dx - \int_{-\infty}^{a} f_X(x)dx = \int_{a}^{b} f_X(x)dx \qquad (2.8)$$

Accordingly, if $F_X(x)$ has a first derivative, then

$$f_X(x) = \frac{dF_X(x)}{dx} \qquad (2.9)$$

It should be noted that any function representing the probability distribution of a random variable must necessarily satisfy the axioms of probability. For this reason, cdf has the following properties: (i) it must be non-negative; (ii) it increases monotonically with the value of the random variable; and (iii) the probabilities associated with all possible values of the random variable must add up to 1.0.

Conditional probability and Bayes' theorem

An important concept in the Bayesian treatment of uncertainties is *conditional probability*. A conditional probability statement is of the following kind:

"*Given the event B (and everything else known is irrelevant for A), then the probability of the event A is r.*"

The above statement is denoted as $P\ (A/B) = r$. Conditional probabilities are essential to a fundamental rule of probability calculus, the product rule. The product rule defines the probability of a conjunction of events:

$$ (2.10) $$

$$P(A, B) = P(A/B)\ P(B)$$

where $P(A, B)$ is the probability of the joint event $A \cup B$, and also called *joint probability distribution* of events A and B. If all these probabilities are conditioned by a context C, then Equation (2.10) can be modified as:

$$P(A, B/C) = P(A/B, C)\ P(B/C) \qquad (2.11)$$

The definition of conditional probability can be used to derive the well known *Bayes' theorem* (Bayes, 1958). From Equation (2.10), it follows that $P(A/B)\ P(B) = P(B/A)\ P(A)$ and after rearranging:

$$P(A/B) = \frac{P(B/A)P(A)}{P(B)} \qquad (2.12)$$

Here $P(A)$ is the *prior* probability of A, that is, the belief in A before the evidence B is considered. $P(B/A)$ is a likelihood of A, which gives the probability of the evidence. $P(A/B)$ is the posterior probability of A which is updated belief in the light of evidence B. $P(B)$, the prior probability of the evidence, is a normalisation constant and is obtained by evaluating the exhaustive and exclusive set of evidence scenario:

$$P(B) = \sum_i P(B/A_i) p(A_i) \tag{2.13}$$

We might intuitively expect that $P(A/B)$ increases with $P(A)$ and $P(B/A)$ according to Bayes' theorem. It provides a direct method for calculating the probability of a hypothesis based on its prior probability, the probabilities of observing various data given the hypothesis, and the observed data itself (Mitchell, 1997).

Measures of uncertainty

Several expressions have been used to describe the degree of uncertainty. The most complete and ideal description of uncertainty is the Probability Density Function (pdf) of the quantity, subject to the uncertainty. However, in most practical problems such a probability function can not be derived or found precisely.

Another measure of the uncertainty of a quantity relates to the possibility to express it in terms of *prediction interval*. The prediction intervals consist of the upper and lower limits between which a future uncertain value of the quantity is expected to lie with a prescribed probability (i.e., confidence level). The endpoints of a prediction interval are known as *the prediction limits*. The width of the prediction interval gives us some idea about how uncertain we are about the uncertain entity.

A useful alternative to quantify the level of uncertainty is to use the statistical moments such as the *variance, standard deviation*, and *coefficient of variation*, when it is difficult to derive or find the pdf.

2.7.2   Fuzzy set theory

In classical set theory, any element x of the universal set X can be classified as being either an element of some sub-set A $(x \in A)$ or an element of its complement $(x \in \overline{A}$, i.e., $x \notin A)$. This is fine if the set A is precisely defined such as "the set of one digit natural numbers". For example, number "9" is member of set A and number "10" is not. There is an abrupt and well defined boundary for numbers "9" and "10" belonging to the set A. However, if set A is a vaguely defined concept like "the set of tall people" then it can be difficult to specify whether a given individual falls into the set or not. In order to overcome this problem, Zadeh (1965) suggested that boundaries of a set should be *fuzzified* so that an element can still be a member of a set with a degree of membership between 0 and 1.

Figure 2.6 shows the example of classical and *fuzzy set* of "tall" people. In Figure 2.6 (a), the height of 1.8 m is precisely defined as tall people (with membership of 1.0), where as the height of 1.6 m is not tall (with membership of 0). However, in the fuzzy set, the height of 1.8 m belongs to the set of tall people with degree of membership 0.95 and the height of 1.6 m belongs to the set of tall people with degree of membership

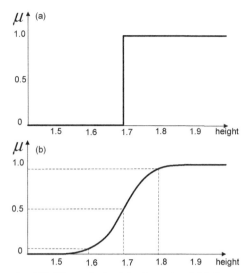

Figure 2.6. An example of "tall" people for (a) crisp set and (b) fuzzy set .

0.05. For a person with height 1.7 m we can say that he/she is either tall or not because he has a membership 0.5 to the set of tall or 0.5 to the set of its complement.

In fuzzy sets, the transition between membership and non-membership can be gradual. This gradual transition of memberships is due to the fact that the boundaries of fuzzy sets are defined imprecisely and vaguely. This property of a fuzzy set makes the *fuzzy set theory* viable for the representation of uncertainty in a non-probabilistic form (Maskey, 2004).

Mathematically the fuzzy set $\tilde{A}$ which is a subset of the universal set X is defined by the *membership function*:

$$\mu_{\tilde{A}}(x): X \to [0,1] \tag{2.14}$$

The membership function $\mu_{\tilde{A}}(x)$ maps every element of the universe of discourse X to the interval [0,1]. Now, fuzzy set $\tilde{A}$ can be defined by ordered pair of the elements of fuzzy set and degree of membership to the set as:

$$\tilde{A} = \{(x, \mu_{\tilde{A}}(x)\} \tag{2.15}$$

where $x \in X$, and $\mu_{\tilde{A}}(x) \in [0,1]$

2.7.3   Entropy theory

The concept of *entropy* plays a pivotal role in information theory. Shannon (1948) is a pioneer who developed entropy theory for the expression of information or uncertainty.

He defined entropy as a quantitative measure of uncertainty associated with a probability distribution or the information content of the distribution. The uncertainty can be quantified with entropy taking into account all the different kinds of available information (Singh, 2000). Thus, entropy is a measure of the amount of uncertainty represented by the probability distribution and is a measure of chaos or the lack of information about a system. Roughly speaking, entropy is a mathematical formulation of the uncertainty and/or the amount of information in a data set. If complete information is available, entropy equals zero, otherwise it is greater than zero.

Small sample size and limited information render estimation of probability distributions of system variables with conventional methods quite difficult. This problem can be alleviated by use of entropy theory, which enables the determination of the least-biased probability distribution with limited knowledge and data. Entropy theory is versatile, robust and efficient (Singh, 1997).

In recent years, entropy theory has been applied to a great variety of problems in hydrology (e.g., Singh and Rajagopal, 1987; Singh, 1997). However, Amorocho and Espildora (1973) were the first to introduce the concept of entropy in the assessment of uncertainty in daily runoff simulation models. Markus et al. (2003) used entropy theory to quantify uncertainty in the results of neural networks based models.

In order to formulate the concept of entropy mathematically, consider an example from Shannon (1948). Suppose we have a set of n possible outcomes x_1, ..., x_n with probabilities of occurrence p_1, ..., p_n respectively. The target is to find a function $H(p_1, ..., p_n)$ which measures the uncertainty in the outcome. The function must satisfy the following conditions:

- H should be continuous in the p_i

- If all the p_i are equal then $p_i = 1/n$. It follows that H should be a monotonically increasing function of n. With equally likely events there is more choice, or uncertainty, when there are more possible events.

- If a choice be broken down into two successive choices, the original H should be the weighted sum of the individual values of H.

The only H satisfying the three above conditions is of the form:

$$H = -k\sum_{i=1}^{n} p_i \log p_i \qquad (2.16)$$

where k is positive constant and depends on the base of logarithm used. Equation (2.16) can be generalized for random variable X with variate values x_n as:

$$H(X) = \sum_{n=1}^{N} P(x_n) \log \frac{1}{P(x_n)} \qquad (2.17)$$

where $P(x_n)$ is the probability of outcome x_n.

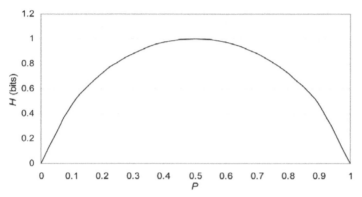

Figure 2.7. Entropy in the case of two possibilities.

Entropy $H(X)$ is also called the *marginal entropy* of a single variable X. It can be shown that the value of $H(X)$ will be maximum when all the x_n are equally likely (i.e. $1/N$). This is also intuitively the most uncertain situation (see Figure 2.7). In this case maximum entropy is given by:

$$H_{\max}(X) = \log N \qquad (2.18)$$

At the other extreme, when all outcomes except one have zero probability, $H(x)$ vanishes; this corresponds to absolute certainty.

2.8 Uncertainty analysis methods

A number of methods have been proposed in the literature to estimate model uncertainty in rainfall-runoff modelling. Through reviews of various methods of uncertainty analysis on rainfall-runoff models can be found in e.g., Melching (1995); Gupta et al. (2005); Montanari (2007); Moradkhani and Sorooshian (2008); Shrestha and Solomatine (2008). Table 2.1 presents references to some of the uncertainty analysis methods used in rainfall-runoff modelling. These methods are broadly classified into six categories:

1. Analytical methods (see, e.g., Tung, 1996);

2. Approximation methods e.g., first-order second moment method (Melching, 1992);

3. Simulation and sampling-based (Monte Carlo) methods (see, e.g., Kuczera and Parent, 1998);

4. Methods from group (3) which are also generally attributed to Bayesian methods e.g., "generalized likelihood uncertainty estimation" (GLUE) by Beven and Binley (1992);

5. Methods based on the analysis of model errors (see, e.g., Montanari and Brath, 2004); and

6. Methods based on fuzzy set theory (see, e.g., Maskey et al., 2004).

All above methods except last one are probabilistic methods. In probabilistic methods, uncertain variables are considered as random and defined by their pdfs. Generally the probabilistic methods involve the propagation of either the pdfs or other statistics of the pdfs through the model.

There are different techniques of uncertainty propagation in probabilistic methods involving different levels of mathematical complexity and data requirements. The appropriate technique to be used depends upon the nature of the problem at hand including the availability of information, model complexity, and type and accuracy of the results desired. The most direct method to assess the uncertainty of a system output is to derive its statistics from knowledge of the statistical properties of the system itself and the input data (Langley, 2000). However, this approach may be limited by two main problems. First, the derivation of the statistics of the output can imply significant mathematical and numerical difficulties; second, the statistical properties of the system and the input may not be known in detail.

Let us consider a function of the form $Y = g(X)$. If the uncertain input X is random, then the output Y is also random and its probability distribution, as well as its moments, will be functionally related to and may be derived from X. For simple problems where the function g is known, it is not so difficult to derive the pdf of the output Y analytically – for example, using the derived distribution method (Ang and Tang, 1975; Tung, 1996).

Table 2.1. References to some of the uncertainty analysis methods used in rainfall-runoff modelling. Note that this list is not exclusive.

Authors	Type of models	Source	Methods
Montanari and Brath (2004)	C	Aggregate	Meta-Gaussian
Shrestha and Solomatine (2006a, 2008)	D, C	Aggregate	Machine learning
Beven and Binley (1992)	P	Parameter	GLUE
Maskey and Guino (2003); Melching (1992)	P	Parameter	FOSM
Kuczera and Parent (1998)	C	Parameter	MCMC
Krzysztofowicz (1999)	P	Aggregate	BFBS
Butts et al. (2004)	P	Structure	
Uhlenbrook et al. (1999)	C	Parameter, Structure	MC

Note - P: Physically based model, C: Conceptual model, D: Data-driven model

However in rainfall-runoff modelling, the functional relationship is so complex and non-linear in multiple variables that analytical derivations are virtually impossible. In such a case sampling based method such as MC simulation or Latin hypercube sampling (McKay et al., 1979) or approximation methods such as Rosenblueth's point estimation (Rosenblueth, 1975), Harr's point estimation (Harr, 1989) or First-Order Second Moment (FOSM) methods (Melching, 1992; Maskey and Guinot, 2003) are inevitable for practical applications. The sampling methods provide the estimation of probability distribution of an output, while approximation methods provide only the moments of the distributions. These methods are described briefly below.

2.8.1 Analytical methods

Derived distribution method also called transformation of variables method is analytical method to determines the distribution of random variable Y through the functional relationship $Y = g(X)$. Given the pdfs or cdfs of the random variable X, the cdf of Y can be obtained as:

$$F_Y(y) = F_X[g^{-1}(y)] \tag{2.19}$$

where $g^{-1}(y)$ represents the inverse function of g. Then the pdf of Y, $f_Y(y)$ can be obtained by taking derivative of $F_Y(y)$ with respect to y as:

$$f_Y(y) = \frac{dF_Y(y)}{dy} = f_X(g^{-1}(y))\frac{dg^{-1}(y)}{dy} \tag{2.20}$$

In principle, the concept of deriving the pdf of a random variable as functions of the pdfs of other random variables is simple and straightforward. However, the success of implementing such procedures largely depends on the functional relationship, the form of the pdfs involved. As mentioned before this method is rarely used in rainfall-runoff modelling, because of the non-linearity and complexity of the model.

2.8.2 Approximation methods

As mentioned before, most of the models used in rainfall-runoff modelling are non-linear and highly complex. This basically limits analytical derivation of the statistical properties of the model output. An alternative approach to estimation and propagation of uncertainty are the "point methods based on first order analysis. The *First Order Second Moment* (FOSM) is one of the most widely used techniques for uncertainty analysis in civil engineering application. The FOSM estimates uncertainty in terms of mean and variance of the system outputs from the given values of mean and variance of input uncertain variables. It uses the first-order terms of the Taylor series expansion about the mean value of each input variable. The great advantage of the FOSM method is its simplicity, computational efficiency, and requiring knowledge of only the first two statistical moments (i.e. mean and variance) of the input variables. Numerical and sometimes even analytical derivatives can be used to calculate the expected value and the variance of the predicted variable (e.g. streamflow). However it has the several theoretical problems. The main weakness is the assumption that a single linearisation of the system performance function at the central values of the basic variables is representative of the statistical properties of system performance over the complete range of basic variables (Melching, 1995). For non-linear system like rainfall-runoff modelling this assumption becomes less and less valid as the basic variables depart from the central values.

Several methods have been proposed to correct the undesirable behaviour of the FOSM method due to linearisation of the system performance function (see, e.g., Melching, 1992; Maskey and Guinot, 2003). Furthermore, other approximation method such as Rosenblueth's point estimation method (Rosenblueth, 1975) also uses the mean and covariance of the variables in a Taylor series expansion, but does not require the calculation of derivatives as FOMS does. Yet another method – Harr's point estimation method (Harr, 1989) reduces the number of simulations required for Rosenblueth's method from 2^p to $2p$, where p is the number of model parameters.

2.8.3 Simulation and sampling-based methods

Monte Carlo (MC) simulation is a sampling method which was named for Monte Carlo, Monaco, where the primary attractions are casinos containing games of chance. The random behaviour in games of chance is similar to how MC simulation selects variable values at random to simulate a model.

MC simulation is an extremely flexible and robust method capable of solving a great variety of problems. In fact, it may be the only method that can estimate the complete probability distribution of the model output for cases with highly non-linear and or complex system relationship (Melching, 1995). It has been used extensively and as a standard means of comparison against other methods for uncertainty assessment.

In MC simulation, random values of each of uncertain variables are generated according to their respective probability distributions and the model is run for each of realizations of uncertain variables. Since we have multiple realizations of outputs from the model, standard statistical technique can be used to estimate the statistical properties (mean, standard deviation etc.) and empirical probability distribution of the model output. The basically MC simulation method involves the following steps (see also Figure 2.8):

1. Randomly sample uncertain variables X_i from their joint probability distributions.

2. Run the model $y = g(x_i)$ with the set of random variables x_i. Store the model output y.

3. Repeat the steps 1 and 2 s times. This gives the realizations of the outputs y_1, ..., y_s.

4. From the realizations y_1, ..., y_s, derive the cdf and other statistical properties (e.g. mean and standard deviation) of Y.

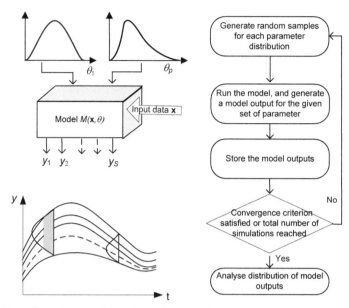

Figure 2.8. Flow chart of Monte Carlo simulation to analyse parameter uncertainty.

The accuracy of the MC method is inversely proportional to the square root of the number of runs s and, therefore, increases gradually with s. As such, the method is computationally expensive, but can reach an arbitrarily level of accuracy. The MC method is generic, invokes fewer assumptions, and requires less user-input than other uncertainty analysis methods. However, the MC method suffers from two major practical limitations: (i) it is difficult to sample the uncertain variables from their joint distribution unless the distribution is well approximated by a multinormal distribution (Kuczera and Parent, 1998); and (ii) It is computationally expensive for complex model.

Markov chain Monte Carlo (MCMC) method such as Metropolis and Hastings (MH) algorithm (Metropolis et al., 1953; Hastings, 1970) has been used to effectively sample parameter from its posterior distribution. In order to reduce the number of samples (model simulations), McKay et al. (1979) introduced efficient sampling scheme so called Latin hypercube sampling.

In a separate line of research within the sampling methods to estimate the parameter uncertainty the following methods can be mentioned: recursive algorithms such as Kalman filter and its extensions ((Kitanidis and Bras, 1980), the DYNIA approach (Wagener et al., 2003), the BaRE approach (Thiemann et al., 2001), the SCEM-UA algorithm (Vrugt et al., 2003). Most of these newly developed methods are based on Bayesian approach.

Most of the probabilistic techniques for uncertainty analysis treat only one source of uncertainty (i.e. parameter uncertainty). Recently attention has been given to other sources of uncertainty, such as input uncertainty or structure uncertainty, as well as integrated approach to combine different sources of uncertainty. The research shows that input or structure uncertainty is more dominant than the parameter uncertainty. For example, Kavetski et al. (2006) and Vrugt et al. (2008a) among others treats input uncertainty in hydrological modelling using Bayesian approach. Butts et al. (2004) analysed impact of the model structure on hydrological modelling uncertainty for streamflow simulation. Recently new schemes have emerged to estimate the combined uncertainties in rainfall-runoff predictions associated with input, parameter and structure uncertainty. For instance, Ajami (2007) used an integrated Bayesian multimodel approach to combine input, parameter and model structure uncertainty. Liu and Gupta (2007) applied integrated data assimilation approach to treat all sources of uncertainty. The readers are referred to the above sources for the detailed descriptions of the methodologies.

Latin hypercube sampling

As mentioned in the previous subsection, MC simulations results inevitably involve sampling errors which decrease as the sample size increases. Increasing sample size for achieving higher precision means an increase in computational time which is not practical in many cases. Variance reduction techniques aim at obtaining high precision for MC simulation results without having to substantially increase the sample size. Several variance reduction techniques are reported in the literature (see, e.g., Tung, 1996); among them *Latin Hypercube Sampling* (LHS) (McKay et al., 1979) is widely used.

LHS is a stratified sampling approach (in contrast to random sampling in MC) that efficiently estimates the statistics of the output. In LHS the probability distribution of each uncertain variable is subdivided into K non-overlapping intervals with an equal probability of $1/K$. A single value is sampled within each interval according to the probability distribution. Thus LHS selects K different values from each of N variables $X_1,\ldots,X_n$. K values obtained from X_1 are paired in a random manner with K values of X_2. These K pairs are combined in a random manner with K values of X_3 to form K triplets, and so on, until K N-tuplets are formed. Thus we create input matrix of size K by N where, i^{th} row consists of specific values of each of N input variables to be used for i^{th} run of the model. The output statistics, probability distribution may then be approximated from the sample of K output values. Example of 5 samples generated by LHS in 2 dimensional spaces is illustrated in Figure 2.9.

Empirical results show that stratified sampling procedure of LHS converges more quickly than random sampling employed in MC and other stratified sampling procedures. Similar to MC, accuracy of LHS is a function of the number of samples K. McKay (1988) suggests that K equals twice the number of uncertain variables might provide a good balance of accuracy and computational cost for models with large numbers of parameters. However, each modeller must check for convergence for the model and problem in question (Melching, 1995). In practice, determining the value of K can be difficult (Helton and Davis, 2003)..

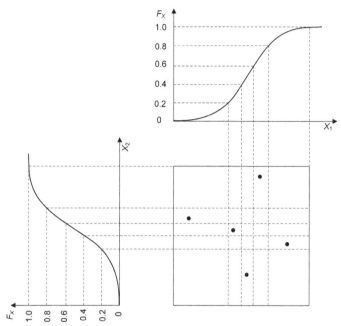

Figure 2.9. Schematic overview of one possible LHS of size 5 for two variables.

Ensemble methods

Ensemble forecast techniques originally deployed in weather forecasts are recently becoming popular in hydrological forecasting as well. These techniques are attractive because they allow effects of a wide range of sources of uncertainty on hydrological forecasts to be accounted for. Forecasting should not only offer an estimate of the most probable future state of the system, but also provide an estimate of the range of possible outcomes. Not only does ensemble forecasting in hydrology offer a general approach to probabilistic forecasting; it offers an approach to improve the hydrological forecasts accuracy as well.

In ensemble method, the deterministic model is run several times with a slightly different initial conditions or parameters. An average, or "ensemble mean", of the different forecasts is created. This ensemble mean will likely have more skill because it averages over the many possible initial states and essentially smoothes the chaotic nature of the system if exists. Based on these ensembles of forecasts, it is possible to forecast hydrological events probabilistically. From the ensembles of the forecasts of the hydrological variables (e.g. runoff), empirical cumulative distribution of the forecasts is derived. The empirical distribution is produced by ranking the M forecasts and calculating the associated probability for each. Let $\{Q_{(1)}, \ldots, Q_{(M)}\}$ denote the order statistics of the ensemble forecasts $\{Q_1, \ldots, Q_M\}$ such that $Q_{(1)} < Q_{(2)} <, \ldots, < Q_{(M)}$. The empirical cumulative distribution is approximated by:

$$P(Q \le Q_{(i)}) = \frac{i}{(M+1)} \qquad (2.21)$$

where P is the probability, i is the rank of the forecast $Q_{(i)}$ in the ensemble, M is the number of ensemble. The main assumptions to derive the probabilistic forecasting from the ensemble forecasts are the following: (i) forecasts are independent realizations of same underlying random process and (ii) the probabilities are computed assuming that each ensemble member is equally likely.

Because of a limited number of realizations, an ensemble forecast does not produce a full forecast probability distribution over all possible events. Furthermore the ensemble forecast underestimated the total uncertainty because not all sources of uncertainty are accounted for in the ensemble generator. Consequently an ensemble forecast does not always constitute a probabilistic forecast (Krzysztofowicz, 2001a).

Ensemble methods are also widely used in data-driven modelling to improve the performance of the resulting model. *Bagging* (Breiman, 1996a; 1996b) and *Boosting* (Schapire, 1990; Freund and Schapire, 1996; 1997) are the two popular ensemble methods that combine the outputs from different predictors to improve the accuracy of prediction. Several studies of boosting and bagging in classification have demonstrated that these techniques are generally more accurate than the individual classifiers (Shrestha and Solomatine, 2006b). The robustness of such techniques is that it is also possible to estimate uncertainty of the prediction.

2.8.4 Bayesian methods

This class of methods combines Bayes' theorem and various variants of MC simulations to either estimate or update the probability distribution function of the uncertain input variables and consequently propagate the distributions from input to the model output. Many of the methods from group (3) are attributed to this class as well. Bayesian forecasting system, GLUE and stochastic emulator based methods are among them and are briefly described in this section.

Bayesian forecasting system

Bayesian approach to uncertainty analysis, named *Bayesian Forecasting System* (BFS) is reported by Krzysztofowicz (1999). The BFS decomposes the total uncertainty about the variable to be forecasted into input uncertainty and hydrologic uncertainty. In the BFS, the input uncertainty is associated with those inputs into the hydrological model which constitute the dominant sources of uncertainty and which therefore are treated as random and has a significant impact on the model output. The hydrologic uncertainty is associated with all other sources of uncertainty such as model, parameter, estimation and measurement errors. The input uncertainty and the hydrologic uncertainty are processed separately through the so-called input uncertainty processor (Kelly and Krzysztofowicz, 2000) and hydrologic uncertainty processor

(Krzysztofowicz and Kelly, 2000) respectively, and integrated by the so-called integrator (Krzysztofowicz, 2001b) to obtain the total uncertainty. The BFS consists of all the mentioned components. For detailed description of the BFBS readers are referred to the citations above and references therein.

GLUE method

GLUE is acronym for *Generalised Likelihood Uncertainty Estimation* introduced by Beven and Binley (1992). The GLUE procedure is a basically Monte Carlo method with constraints on the admissible parameters vectors, and is based on the premise that there are many different model structures and many different parameter sets within chosen model structure that may be behavioural or acceptable in reproducing the observed behaviour of the system. This concept is called *equifinality* (Beven and Freer, 2001). Thus, the GLUE procedure recognizes that no single optimum parameter set exists in the calibration of distributed model, but rather a range of different sets of model parameter values may represent the process equally well. Savenije (2001) argue that equifinality is a blessing rather than a curse; it is in fact the justification for many of the hydrological theories we use.

The GLUE procedure consists of making a large number of runs of a given model with different sets of parameter values chosen randomly from specified parameter distributions. The output from each model run is compared to the corresponding observed data and each model run (associated with a set of parameter values sampled) is assigned a likelihood value. The likelihood value is related to the performance of the model (a way to measure how well the model fits the observed data) and thus, higher values of likelihood typically indicate better correspondence between the model outputs and observations. All simulations (i.e., model runs) with likelihood measures greater than predefined threshold value are considered behavioural and retained for further analysis. The likelihood values of the behavioural simulations are rescaled such that the sum of the likelihood values equal unity. Finally, the uncertainty of the model output is estimated by deriving the likelihood weighted cumulative distribution function of the output from the behavioural simulations.

The GLUE methodology also recognises that as more data or different types of data are made available, it may be necessary to update the likelihood measures associated with different simulations. This is achieved quite easily using Bayes' theorem, which allows a prior distribution of likelihood weights to be modified by a set of likelihood weights arising from the simulation of a new data set to produce an updated or posterior likelihood distribution (Freer et al., 1996). This updating of likelihood weight is important to refine the uncertainty estimate over time as more observed data become available. The details of the GLUE method can be found elsewhere (e.g., Beven and Binley, 1992; Freer et al., 1996; Beven and Freer, 2001; Beven, 2006).

GLUE is one of the popular methods in analysing the parameter uncertainty in hydrological modelling and widely used over the past ten years to analyse and estimate predictive uncertainty, particularly in hydrological applications (see, e.g., Freer et al., 1996; Beven and Freer, 2001; Montanari, 2005). Users of GLUE are attracted by its simple understandable ideas, relative ease of implementation and use, and its ability to handle different error structures and models without major modifications of the method itself. Despite its popularity, there are theoretical and practical issues related to GLUE method reported in the literatures. For instance, Mantovan and Todini (2006) argue that GLUE is inconsistent with the Bayesian inference processes that leads to overestimation of uncertainty, both for the parameter uncertainty estimation and predictive uncertainty estimation. For this the readers are referred to Mantovan and Todini (2006) and the subsequent discussions in the Journal of Hydrology in 2007 and 2008.

A practical problem with the GLUE method is that, for models with a large number of parameters, the sample size from the respective parameter distributions must be very large to achieve a reliable estimate of model uncertainties (Kuczera and Parent, 1998). In order to reduce the number of sample size (i.e., model runs) hybrid genetic algorithm and artificial neural network is applied by Khu and Werner (2003). Blasone et al. (2008) used adaptive Markov chain Monte Carlo sampling within the GLUE methodology to improve the sampling of the high probability density region of the parameter space. Yet another way to reduce the number of Monte Carlo simulations could be to use SCEM-UA algorithm (Vrugt et al., 2003).

Another practical issue is that the percentage of the observations falling within the prediction limits provided by GLUE are much lower than the given confidence level used to produce these prediction limits in many cases (see. e.g., Montanari, 2005). Xiong and O'Connor (2008) modified the GLUE method to improve the efficiency of the GLUE prediction limits in enveloping the observed discharge.

Stochastic emulator based method

As mentioned before, the standard techniques (i.e. MC simulation) for uncertainty analysis demand a very large number of model runs, and when a single model run takes several minutes these methods become impractical. Over the last 10 years, a range of efficient tools have been developed using Bayesian statistics for calibration, sensitivity and uncertainty analysis of complex models. One of the tools is the use of emulator, which is a statistical approximation of the simulator (O'Hagan, 2006). If the approximation is good enough, then the uncertainty analysis produced by the emulator will be sufficiently close to those that would have been obtained using the original simulator.

Let function $f(.)$ is the simulator that maps inputs $\mathbf{x}$ into an scalar output $y = f(\mathbf{x})$ and $\hat{f}(.)$ be the approximation of it. An emulator not only provides an approximation to $f(\mathbf{x})$ but also an entire probability distribution for $f(\mathbf{x})$. The two most desirable properties of the emulator $\hat{f}(.)$ for uncertainty analysis are: (i) $\hat{f}(.)$ should be simpler than the original function $f(.)$; and (ii) $\hat{f}(.)$ should be computed much faster than $f(.)$. The description of building an emulator can be described briefly as follows:

1. Construct training data by running the simulator for different configuration of input data $\mathbf{x}$. Here input data $\mathbf{x}$ represents all uncertain variables (e.g. forcing data, model parameters or initial condition etc) of the simulator.

2. Consider the simulator $f(.)$ as unknown random function and represent it by suitable statistical function $\hat{f}(.)$. One of the functions is Gaussian process (see, e.g., Kennedy and O'Hagan, 2001; O'Hagan, 2006; Goldstein and Rougier, 2009) which is more efficient and analytically very tractable. A Gaussian process is an extension of normal or Gaussian distribution. It is a distribution for a function, where each point $f(\mathbf{x})$ has a normal distribution.

3. Assume prior distribution for the hyper-parameters of the Gaussian process. The posterior distribution of the hyper-parameters is estimated using the training data. This posterior distribution is the emulator.

The detailed description of the Gaussian process emulator and its application can be found in the citation above.

2.8.5   Methods based on the analysis of model errors

Meta-Gaussian method

The meta-Gaussian method (Montanari and Brath, 2004) is a statistical method to estimate the uncertainty by analyzing the model residuals that occurred in reproducing the observed historical data. This method is based on meta-Gaussian bivariate probability distribution model introduced in hydrology by Kelly and Krzysztofowicz (1997). The bivariate meta-Gaussian distribution is constructed by fitting with a bivariate Gaussian distribution random variables with arbitrary marginal distribution by embedding the normal quantile transform (NQT) of each variate into the Gaussian law.

The meta-Gaussian method computes model uncertainty by estimating the pdf of the model error conditioned by the contemporary value of the simulated river flow. In this method, both model residuals and simulated model outputs are transformed into the Gaussian domain by NQT. Let $S(t)$ and $E(t)$ denote the stochastic processes of the model simulation s_t and the corresponding error e_t, respectively. The arbitrary marginal distribution of $S(t)$ and $E(t)$ are denoted by $P(S \leq s_t)$ and $P(E \leq e_t)$. The procedure of the meta-Gaussian method can be described briefly as follows:

1. NQT is applied in order to make $P(S \leq s_t)$ and $P(E \leq e_t)$ Gaussian. The NQT involves the following steps: (i) the cumulative frequency $F(s_t)$ is computed for each s_t using the Weibull plotting position, that is; $F(s_t) = j_t/(n+1)$. Here j_t is the position occupied by s_t in the data rearranged in ascending order. (ii) The standard normal quantile Ns_t for each $F(s_t)$ is computed using:

$$NS(t) = Q^{-1}[P(S \le s_t)] \tag{2.22}$$

where Q^{-1} is inverse of the standard normal distribution, and distribution $P(S \le s_t)$ of each s_t is approximated with corresponding sample frequency $F(s_t)$.

2. The discrete mapping of s_t in its normalised space Ns_t is obtained using equation (2.22). This mapping is used to interpolate linearly the inverse of Ns_t to get back in original space.

3. The cross dependence between $NS(t)$ and $NE(t)$ is computed according to the formula:

$$Ne_t = \sigma Ns_t + N\varepsilon_t \tag{2.23}$$

where σ is the linear correlation coefficient between $NS(t)$ and $NE(t)$, and $N\varepsilon_t$ is the outcome of the stochastic process.

4. The conditional mean μ and variance σ^2 of the model error in the normalise space $NE(t)$ are given by:

$$\mu = \rho Ns_t \tag{2.24}$$
$$\sigma^2 = 1 - \rho^2$$

5. The prediction limits corresponding to α significance level is computed according to the formula:

$$PL^U = s_t + Q^{-1}[\mu + z_{\alpha/2}\sigma] \tag{2.25}$$
$$PL^L = s_t + Q^{-1}[\mu - z_{\alpha/2}\sigma]$$

where PL^U and PL^L are the upper and lower prediction limits, respectively, $z_{\alpha/2}$ is the value of the standard normal variate with cumulative probability level of $\alpha/2$.

The meta-Gaussian approach is based on certain assumptions about the residuals of the normal linear equation (i.e., equation (2.23)): they should be Gaussian and homoscedastic. Furthermore, the regression of $NE(t)$ on $NS(t)$ should be linear. However, in practical application, some of these basic assumptions are not satisfied. It is suggested to verify the goodness of the fit provided by the meta-Gaussian model. The lack of fit can be resolved by transforming the model residuals according to the outline presented by Montanari and Brath (2004) to stabilize their variances. Recently Montanari and Grossi (2008) used separate meta-Gaussian models for positive and negative errors if good fit was not achieved through the linear regression. The details of the meta-Gaussian method and its applications to estimate the uncertainty of rainfall-runoff simulations can be found in the citation above.

Quantile regression method

Much less known quantile regression (QR) introduced by Koenker and Bassett (1978) is a statistical technique intended to estimate the conditional quantile functions. Just as classical linear regression methods based on minimizing sums of squared residuals enable one to estimate models for conditional mean of the response variable given certain values of the predictor variables, QR methods offer a mechanism for estimating models for the conditional median function, and the full range of other conditional quantile functions. By supplementing the estimation of conditional mean functions with techniques for estimating an entire family of conditional quantile functions, QR is capable of providing a more complete statistical analysis of the stochastic relationships among random variables.

The *p*th regression quantile $(0 < p < 1)$ for output variable y is defined as any solution to the minimization problem:

$$\min_{b \in \Re^K} \left[\sum_{t \in \{t: y_t \geq x_t b\}} p|y_t - x_t b| + \sum_{t \in \{t: y_t < x_t b\}} (1-p)|y_t - x_t b| \right] \tag{2.26}$$

where x_t is the vector of input, b is the solutions of equation. This minimization problem can be solved very efficiently by linear programming methods.

Although the QR technique was applied successfully in many fields because of its insensitivity to outliers and avoidance of parametric assumptions, its application to compute quantiles of the simulations of the conceptual or distributed rainfall-runoff model is not straightforward because of the complexity and non-linearity of the rainfall-runoff process which hinder to present inputs and output of the rainfall-runoff model in the form of equation (2.26). However, it is possible to use it to predict the conditional quantiles of the model errors of runoff predictions. In this thesis we have also used QR techniques to estimate the quantiles of the rainfall-runoff model errors.

2.8.6 Fuzzy set theory-based methods

Fuzzy set theory provides a means for representing uncertainties. It is powerful tool for modelling the kind of uncertainty associated with vagueness, with imprecision, and/or with a lack of information regarding a particular element of the problem at hand. The underlying power of fuzzy set theory is that it uses linguistic variables, rather than quantitative variables to represents imprecise concepts. Fuzzy extension principle based method and fuzzy α-cut technique are the widely used fuzzy set theory based method for uncertainty handling. The *fuzzy extension principle* provides a mechanism for the mapping of the uncertain input variables defined by their membership functions to the resulting uncertain output variable in the form of a membership function (Maskey, 2004).

Fuzzy α-cut technique is approximation for fuzzy extension principle method as direct application of the extension principle is very time consuming and difficult for complex problems involving many input variables. An α-cut is a crisp set consisting of elements of $\tilde{A}$ which belongs to the fuzzy set at least to degree α. An example of α-cut with triangular membership function is shown in Figure 2.10. The α- cut level intersects at two points a and b on the membership function. The values of the variables x corresponding to the points a and b are x_1 and x_2 ($x_1, x_2 \in X$) respectively. Then the set A_α contains all the possible values of the variables X including and between x_1 and x_2. The x_1 and x_2 are commonly referred as lower and upper bounds of the α-cut, respectively. By considering the fuzzy variable at a given α- cut level, operations on fuzzy set can be reduced to operations within the interval arithmetic.

Fuzzy set theory-based methods are powerful but they also have some limitations for modelling uncertainty:

- They cannot handle all type of uncertainty but suitable only for a kind of uncertainty coming from vagueness;

- They require knowledge of the membership function of the uncertain variables which is very subjective;

- They cannot incorporate the effect of the correlation between the uncertain input variables.

The details of the method can be found in Maskey (2004) and its application in hydrological modelling is reported in Maskey et al. (2004) and in Abebe et al. (2000).

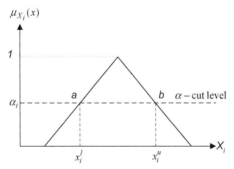

Figure 2.10. Illustration of an α -cut of fuzzy set.

Chapter 3
Machine Learning Techniques

This chapter presents an overview of the machine learning techniques used in this research. Brief descriptions of machine learning techniques namely artificial neural networks, model trees, instance based learning, and some clustering techniques are given. These learning methods are extensively used in the following chapters in modelling the uncertainty of the rainfall-runoff models.

3.1 Introduction

In Chapter 2, we presented three classes of rainfall-runoff models and introduced data-driven modelling. Data-driven modelling utilises the theoretical foundations and methods of machine learning applied to the measured data (Solomatine, 2002). Machine learning is a subfield of artificial intelligence that is concerned with the design and development of algorithms that allow computers (machines) to improve their performance over time, based on data. A major focus of machine learning research is to automatically produce (induce) models from data through experience.

Learning from data in machine learning is not a completely new phenomenon. Statistical methods, known for a long time, also learn from data. However, in most of these methods the structure of the model to be built has to be determined first, which is done with the aid of the empirical or analytical approaches, before the unknown model parameters can be estimated (Chakraborty et al., 1992). In this sense these methods can be seen as the model-driven approaches. The greatest plus of a model-driven approach is that it produces a simple and understandable picture of the relationship between the input variables and the responses or output (Breiman, 2001). However, in model driven approaches, the model that generates the data is assumed or should be known priori, and in many statistical models strong assumptions about the properties of data (e.g., normality) are made.

In contrast, most machine learning is positioned as the methods learning from the data or examples and able to capture subtle functional relationship between the input variables and the output, even if the underlying mechanism producing data is not known or hard to describe. This modelling approach with the ability to learn from experience is very useful for many practical problems since it is easier to have data than to have good theoretical guesses about the underlying laws governing the system from which data are generated. Note however, that there is no clear division between statistical methods (some of which are called statistical learning methods) and machine learning (Hastie et al., 2001)

Machine learning is a multidisciplinary field which is enriched with the concepts drawn from many fields such as statistics, artificial intelligence, information technology, neurobiology, physiology, control theory and other. With ever growing computational power and development of new techniques to acquisition, storage and transmission of data, machine learning methods bloomed dramatically in the last decade. In recent years many machine learning techniques have been applied successfully in many fields including finance, autonomous vehicle driving on public highways, robotics, military applications, engineering, etc.

3.2   Machine learning types

There are basically three types of machine learning methods, namely supervised, unsupervised and reinforcement learning. Supervised learning is a machine learning technique for learning functional relationships between the input data and the output from training data. The training data consist of pairs of input data and desired outputs. The output of the function can be a continuous value (in this case the style of learning is called regression), or can predict a class label of the input variable (style of learning is called classification). The task of a supervised learner is to predict the output for any new input vector after having seen a number of training examples. To achieve this, the learner has to generalize from the presented data to unseen situations with a reasonable accuracy. The examples of supervised learning methods are artificial neural networks, decision trees, model trees, instance based learning, support vector machines, etc.

In the unsupervised learning, the training data consists of only input data and the target output is not known. The learner task is to determine how the data are organized and to reveal some important features or characteristics of the data. One form of unsupervised learning is clustering, where the data are grouped into subsets according to similarity between them.

Reinforcement learning is machine learning approach concerned with how an autonomous agent that senses and acts in its environment can learn to choose optimal actions to achieve its goal (Mitchell, 1997). Reinforcement learning algorithms attempt to find a policy that maps states of the world to the actions the agent ought to take in those states. They incorporate learning a mapping from situations to actions by trial and error interactions with a dynamic environment. The components of reinforcement learning are: an agent, a dynamic environment to act, and a reward function. Reinforcement learning differs from the supervised learning problem in that correct input/output pairs are never presented. Reinforcement learning is used in modelling an optimal control strategy.

Machine learning methods can be also classified into eager learning or lazy learning. The former type constructs explicitly the model to predict or classify the output as soon as the training examples are presented. The latter type simply stores the presented training data and postpones using it until a new instance or input has to be predicted or classified. Examples of eager learning are artificial neural networks, decision trees and model trees etc. K-nearest neighbour and locally weighted regressions are lazy learning methods.

In this chapter supervise learning techniques - artificial neural networks and model tress; lazy learning technique – locally weighted regression; and unsupervised learning technique – clustering methods are described briefly.

3.3   Learning principle and notations

Before going to the details of the principles of machine learning, it is required to define some terminologies which are frequently used in the rest of the thesis. Consider a set of n examples $D = \{X, \mathbf{y}\}$ consisting of p input variables and one dimensional output. Here data D is a set of input-output pairs. $X = \{\mathbf{x}_1, \ldots, \mathbf{x}_p\}$ is the matrix of the input where $\mathbf{x}_j$ is the column vector having n number of examples and p is the number of variables. $\mathbf{y}$ is the vector of the desired output. Hence data D is n by $(p+1)$ matrix, where input data matrix X is size of n by p. Note that $\mathbf{x} = \{x_1, \ldots, x_p\}$ represents a single input vector and y represents the corresponding scalar output.

Instance, example or *a data point* is used to describe a record of a data and these terms are used interchangeably thought the thesis. Likewise the components of input vector are variously called attributes, input variables or features and used interchangeably. The term *output* is used in conjunction with the terms like "observed" or "measured" and thus *observed* (or *measured*) *output* denotes the system's output variable which has to be predicted from the model. The response from the model is termed as *model output* or simply *output* or *prediction*. In this thesis *machine learning model* is also called simply the *model*.

Unless specified, the following indices and notations are used in this thesis - i: index for a row of data, n: total number of data, j: index for a column of data, p: number of the attributes, k: index for the output, l: number of the output variables.

Having defined some terminology, we can now present the principle of machine learning in the context of supervised learning. A machine learning technique, on the basis of observed data D tries to identify ("learn") the target function $f(\mathbf{x},\mathbf{w})$ describing how the real system behaves. In other words, the learning aim is to approximate function f that operates on input data $\mathbf{x}$ to predict the output y with a desired degree of accuracy. Learning (or *training*) here is the process of minimizing the difference between observed response y and model response $\hat{y}$ through the optimisation procedure. Learning is an iterative procedure starting from the initial guess of the parameter vector $\mathbf{w}$ and updating it by comparing the models' response $\hat{y}$ with the

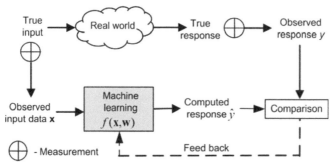

Figure 3.1. Machine learning algorithm using observation of a physical system $D = \{X, \mathbf{y}\}$ to learn an unknown mapping function $f(\mathbf{x}, \mathbf{w})$. During the learning phase (training), the prediction error $y - \hat{y}$ is used to update the parameter vector $\mathbf{w}$ and/or function f.

observed response y. Thus in each iteration, the error function, generally the sum of squared errors, is computed and its value is used to update the parameter vector $\mathbf{w}$. The learning process is continued until such time as an acceptable degree of accuracy is achieved (however there are other criteria to terminate training, e.g. using the cross-validation set, but they are skipped here for simplicity). Once the learning process is completed, it is assumed that the underlying function f that generated the data has been discovered (or rather its approximation). This learning process is presented in Figure 3.1. Please note that observed input and response are contaminated by the measurement errors. Hence the true function that generated the data is never known and cannot be approximated by the limited data and possibly contaminated with measurement errors.

Generally, learning is done only on the subset of data called training data and the rest of the data is used to check the generalization capability of the trained model. In other words, the optimal parameter vector $\mathbf{w}$ of the model is determined using the training data set, while its predictive performance is evaluated by presenting the unseen input data. The unseen or independent input data is called test (or verification or validation) data. It is important to note that the test data should not be used during the training process in any capacity.

If the difference in the error obtained using the test set is markedly different than that obtained using the training data, then it is likely that the two data sets are not representative of the same population, or that the model overfits the data. It is observed especially in cases where learning is performed too long or where training examples are rare, so that the learner may adjust to very specific random features or noises of the training data, which have no causal relation to the output. This phenomenon of fitting to the noise in the training data is called overfitting. In the process of overfitting, the performance on the training examples still increases while the performance on unseen data becomes worse. Hence overfitting should be avoided in the training process. There are different methods to minimise the overfitting in different learning methods, however one which is common to all learning method is to use the third data set, apart from the training and test data sets. This third data set is often called cross-validation data set. The error is monitored in the cross-validation data set while training the model in the training data set. The training process is stopped when the error in the cross-validation data begins to rise even when the training error is still decreasing.

3.4   Application of machine learning in rainfall-runoff modelling

Over the last decade many machine learning techniques have been used extensively in the field of rainfall-runoff modelling. Among them the artificial neural networks (ANNs) is most popular and widely used technique. Maier and Dandy (2000) provides an extensive reviews of ANNs application which was published in international journal until the end of 1998 (e.g., Hsu et al., 1995 ; Minns and Hall, 1996; Dawson and Wilby, 1998). Recent application of ANN in rainfall-runoff modelling can be found elsewhere (e.g., Govindaraju and Rao, 2000; Abrahart et al., 2004). Apart from ANN, other machine learning techniques have been also used, e.g. model trees (Solomatine and Dulal, 2003), fuzzy rules based system (Bárdossy and Duckstein, 1995), support vector machines (Dibike et al., 2001; Bray and Han, 2004) genetic programming (Whigham and Crapper, 2001; Liong et al., 2002).

Machine learning techniques have been also used to improve the accuracy of prediction/forecasting made by process-based rainfall-runoff model. Generally they are used to update the output variables by forecasting the error of the process based models. Shamseldin and O'Connor (2001) and Xiong and O'Connor (2002) used ANNs to update runoff forecasts of the soil moisture accounting and routing (SMAR) conceptual model. Brath et al. (2002) used ANNs and nearest neighbor methods to update the runoff forecasting issued by a ADM conceptual rainfall-runoff model (Franchini, 1996). Abebe and Price (2003) used ANNs to forecast errors of ADM conceptual model (Franchini, 1996) and they viewed this approach as a complementary modelling. Recently Corzo et al. (2009) combined semi-distributed process-based and data-driven models in flow simulation.

Besides this, machine learning techniques have been also employed to improve the performance of the data-driven rainfall-runoff models (e.g., Xiong et al., 2004). All these techniques to update the model predictions can be seen as error modelling paradigm to reduce the uncertainty of the rainfall-runoff model. However these techniques do not provide explicitly the uncertainty of the model prediction.

Recently, Shrestha and Solomatine (2006a, 2008), Solomatine and Shrestha (2009), and Shrestha et al. (2009) developed novel methods to estimate uncertainty of the runoff prediction by rainfall-runoff models using machine learning techniques. These methods are the original contribution of the present research and described in the remaining chapters.

3.5   Artificial neural networks

An *artificial neural network* (ANN) is an information processing paradigm that is inspired by the way biological nervous systems, such as the brain, process information. A neural network is massively parallel-distributed processor made up of simple processing units, which has a natural propensity for storing experiential knowledge and making it available for use. It resembles the human brain in the following two ways: a neural network acquires knowledge through learning, and a neural network's knowledge is stored within inter-neuron connection strengths known as synaptic weights (Haykin, 1999). ANNs are nowadays regarded as universal function approximation due to their ability to represent both linear and complex non-linear relationships and in their ability to learn these relationships directly from the data being modelled. They provide a robust approach to approximating real-valued and, after certain modification, discrete-valued functions. Although the concept of artificial neuron was first introduced by McCulloch and Pitts (1943), research into application of ANNs has blossomed since the introduction of the backpropogation training algorithm for feedforward ANNs. It was suggested by P. Werbos in 1974 (Werbos, 1974)), but that became widely known only after publication of the edited 2-volume work by Rumelhart et al. (1986). Nowadays they have been applied successfully in many fields for control, classification, pattern recognition, modelling dynamic systems and time series forecasting.

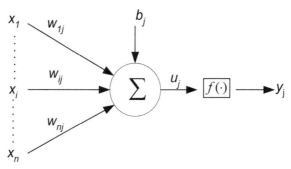

Figure 3.2. Schematic diagram of an artificial neuron.

ANNs consist of a large number of simple processing elements called *neurons* or *nodes*. Each neuron is then connected to other neurons by means of direct links, each being associated with a weight that represents information being used by the network in its effort to solve the problem. The neural network can be in general characterised by architecture (the patterns of connection between the neurons), training or learning algorithm (the methods of determining the weights on the connections) and activation function. The architecture of a typical neural network with a single neuron is shown in Figure 3.2. It consists of five basic elements:

1. Input nodes for receiving input signals x_1, …, x_p;

2. A set of connecting links (often called *synapses*), each of which is characterised by a weight w_{ij};

3. Aggregating function to sum the input signals;

4. Activation function that calculates the activation level of the neuron; and

5. Output nodes $y_1, ..., y_l$

The processing of each neuron is carried out in two steps: (i) summing of the weighted input signals, and (ii) applying activation function to the sum for limiting the amplitude of the output of a neuron. Mathematically it can be described by the following two equations:

$$u_j = \sum_{i=1}^{p} w_{ij} x_i \qquad (3.1)$$

$$y_j = f(u_j + b_j) \qquad (3.2)$$

where w_{ij} is the weight connecting the input i to the neuron j. The effective incoming signal u_j and bias b_j is passed through activation function $f(.)$ to produce the output signal y_j. The main difference between the neurons in common use lies in the type of the activation function. Their functional form determines the response of a node to the total input signal, however there is one thing common among these activation functions – they all restrict the input signals to certain limits. Some commonly used activation functions are linear, binary, sigmoid and tangent hyperbolic.

3.5.1   Learning in ANN

In order to generate an output vector $\mathbf{y} = (y_1, ..., y_n)$ from an ANN that is as close as possible to the target vector $\mathbf{t} = (t_1, ..., t_n)$, a training (learning) process, also called learning is employed to find optimal weight matrices W and bias vectors $\mathbf{b}$, that minimise a predetermined error function usually sum squared error that has the form:

$$J = \frac{1}{2} \sum_{k=1}^{l} \sum_{i=1}^{n} e_{ik}^{2} \qquad (3.3)$$

where l is the number of output nodes, e_{ik} is the kth output error for the ith example. The error function of equation (3.3) is the aggregation of the errors over all of the output nodes. The factor ½ in equation (3.3) appears when the assumption of normality of model errors, traditional for statistical models, is made (actually not used here); it also convenient to leave it to simplify the subsequent derivations resulting from the minimization of J with respect to the weights of the network. Several algorithms are known to minimise the error function J based on the error signal e_{ik}, such that the actual response of each output node in the network is close to the desired response for that neuron. Gradient descent is one of such algorithms to determine a weight vector that minimizes J by starting with an arbitrary initial weight vector, and then repeatedly updating it in small steps by computing the error function. At each step, the weight vector is altered in the direction that produces the steepest descent along the error surface. This process continues until the minimum (or small enough) error is reached, or until a predefined number of steps are made.

The direction of the steepest descent along the error surface can be found by computing the gradient of the error surface. The gradient of the error surface is a vector that always point toward the direction of the maximum J change and with a magnitude equal to the slope of the tangent of the error surface. The training rule for the gradient descent is given by:

$$\mathbf{w}(iter+1) = \mathbf{w}(iter) - \eta \nabla J(iter) \tag{3.4}$$

where $iter$ is iteration number, η is a positive constant called learning rate which determines the step size in the search, $\nabla J(iter)$ denotes the gradient of the error surface at the iteration $iter$. Traditionally the difference operator is used to estimate the gradient which is straightforward but not very practicable. Fortunately, an elegant approach to estimate the gradient was proposed as follow:

$$\nabla J(iter) = -e(iter)x(iter) \tag{3.5}$$

The equation (3.5) tells us that an instantaneous estimate of the gradient at iteration $iter$ is simply the product of the current input to the weight times the current error. Combining equation (3.4) and (3.5), the steepest descent equation becomes:

$$\mathbf{w}(iter+1) = \mathbf{w}(iter) + \eta e(iter)x(iter) \tag{3.6}$$

3.5.2 Multi-layer perceptron network

Multi-Layer Perceptron (MLP) is one of the most popular and successful neural network architectures. The MLP consists of an input layer, an output layers and at least one intermediate layer between input and output layer. The first layer is the input layer, which receives the input signal. The intermediate layers are known as hidden layers, which do not have direct connection to the outer world. The last layer is the output layer at which the overall mapping of the network input is made available and thus represents the model output. The nodes in one layer are connected to those in the next, but not to those in the same layer. Thus, the information or signal flow in the network is restricted to a flow, layer by layer, from the input to output through hidden layers. Figure 3.3 shows the example of a three layer MLP with one hidden layer.

Training (i.e. computing the weights) of the MLP networks is done with the back-propagation algorithm which is the most popular algorithm capable of capturing a variety of non-linear error surfaces. It is essentially a gradient descent technique that minimises the network error function. The back propagation algorithm involves two steps: the first step is feed forward pass, in which the input vectors of all the training

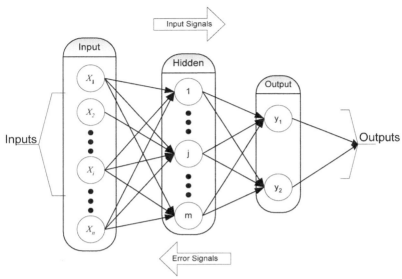

Figure 3.3. A three layer MLP architecture.

examples are fed to the network and the output vectors are calculated. The performance of the network is evaluated using an error function that is based on the target and network outputs. After the error is computed, the back propagation step starts, in which the error is propagated back to adjust the network weights and bias. The iteration continues until the outputs of the network will match with the targets with the desired degree of accuracy. In a typical application, the weight update loop in back propagation may be iterated thousands of times. A variety of termination condition can be used to halt the iteration. One may choose to halt after a fixed number of iterations or once the error on the training examples falls below some threshold or once the error on separate validation set of examples meets some criterion. The choice of termination or stopping criterion is important and discussed by many authors (see, e.g., Mitchell, 1997).

Although the back propagation method does not guarantee convergence to an optimal solution since local minima may exist, it appears in practice that it leads to solutions in almost every case. In fact, standard multi-layer, feed-forward networks with only one hidden layer have been found capable of approximating any measurable function to any desired degree of accuracy. Detailed description of the back propagation algorithm can be found in Haykin (1999), Principe et al. (1999) among others.

3.6   Model trees

Model trees (or M5 model trees) are relatively new machine learning technique introduced by Quinlan (1992) who also suggested the algorithm that uses information theory to build them – the M5 algorithm. This is effectively a piece-wise linear regression model. A complex modelling problem can be solved by dividing it into a number of simple tasks and building simple model for each of them. In other words, the data set is divided into a number of subsets for each of which separate specialized model is built. These models are often called *experts* or *modules*; and the combination of experts is called a modular model, or a *committee machine*.

There are several ways to classify a committee machine (see, e.g., Haykin, 1999; Solomatine and Siek, 2006). Committee machines can be often classified based on how the input space is split. According to this way, it is possible to classify committee machine into four categories (Solomatine, 2005): (i) hard splitting where individual experts are trained on particular subspaces of data independently and use only one of them to predict the output for a new input vector, e.g. model trees, regression trees (Breiman et al., 1984); (ii) hard splitting – soft combination where outputs are combined by 'soft' weighting scheme; (iii) statistically-driven soft splitting, used in mixtures of experts (Jacobs et al., 1991) and boosting schemes (Freund and Schapire, 1996; Shrestha and Solomatine, 2006b); and (iv) no-split option leading to ensembles; the models are trained on the whole training set and their outputs are combined using a weighting scheme where the model weights typically depend on model accuracy, e.g. bagging (Breiman, 1996).

A model tree (MT), as mentioned previously, belongs to a class of committee machine which uses the 'hard' (i.e. yes-no) splits of input space into regions progressively narrowing the regions of the input space. Thus model tree is a hierarchical (or tree-like) modular model which has splitting rules in non-terminal nodes and the expert models at the leaves of the tree. In M5 model trees the expert models are often simple linear regression equation derived by fitting to the non-intersecting data subsets. Once the expert models are formed recursively in the leaves of the hierarchical tree, then prediction with the new input vector consists of the two steps: (i) classifying the input vector to one of the subspace by following the tree; and (ii) running the corresponding expert model. Brief description of model tree algorithm is presented below.

The M5 algorithm for inducing a model tree was developed by Quinlan (1992). Assume we are given a set of N data pairs $\{x_i, y_i\}$, $i = 1, \ldots, n$, denoted by D. Here x is p dimensional input vector (i.e. $x^1, \ldots, x^p$) and y is target. Thus, a pair of input vector and target value constitutes the example, and the aim of the building model tree is to map the input vector to the corresponding target by generating simple linear equations at the leaves of the trees. The first step in building a model tree is to determine which input variable (often called attribute) is the best to split the training D. The splitting criterion (i.e. selection of the input variable and splitting value of the input variable) is based on treating the standard deviation of the target values that reach a node as a measure of the error at that node, and calculating the expected reduction in error as a result of testing each input variable at that node. The expected error reduction, which is called standard deviation reduction, SDR, is calculated by

$$\text{SDR} = sd(T) - \sum_i \frac{|T_i|}{|T|} sd(T_i) \qquad (3.7)$$

where, T represents set of examples that reach the splitting node, T_1, T_2,..., represents the subset of T that results from splitting the node according to the chosen input variable, sd represents standard deviation, $|T_i|/|T|$ is the weight that represents the fraction of the examples belonging to subset T_i.

After examining all possible splits, M5 chooses the one that maximises SDR. The splitting of the training examples is done recursively to the subsets. The splitting process terminates when the target values of all the examples that reach a node vary only slightly, or only a few instances remain. This relentless division often produces over-elaborate structures that must be pruned back, for instance by replacing a subtree with a leaf. In the final stage, 'smoothing' is performed to compensate for the sharp discontinuities that will inevitably occur between the adjacent linear models at the leaves of the pruned tree. In smoothing, the outputs from adjacent linear equations are updated in such a way that their difference for the neighboring input vectors belonging to the different leaf models will be smaller. Details of the pruning and smoothing process can be found in Witten and Frank (2000). Figure 3.4 presents an example of model tree.

As compared to other machine learning techniques, model tree learns efficiently and can tackle tasks with very high dimensionality—up to hundreds of variables. The main advantage of model tree is that results are transparent and interpretable. Recently two other versions of M5 algorithm were proposed by Solomatine and Siek (2006): M5opt that allows for deeper optimization of trees, and the M5flex that gives a modeller more possibilities to decide about how the data space is split in the process of building regression models. During the last years several authors have shown the effectiveness of the M5 machine learning method in rainfall-runoff modelling (see, e.g., Solomatine and Dulal, 2003; Solomatine and Siek, 2006; Stravs and Brilly, 2007).

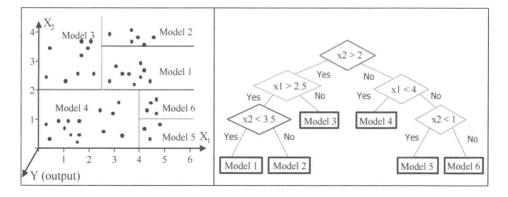

Figure 3.4. Example of model trees.

3.7 Instance based learning

Many machine learning methods including those described in sections 3.5 and 3.6 are model-based methods meaning that they construct explicitly the model to predict or classify the output as soon as the training examples are provided. After training, the model is used for predictions and the data are generally discarded. In contrast, memory-based method - instance based learning (IBL) simply stores the presented training data and postpone using it until a new instance or input has to be predicted or classified. When a new input vector is presented to the model, a subset of similar instances is retrieved from the previously stored examples and their corresponding outputs are used to predict or classify the output for the new query vector (instance). IBL methods, in fact, construct a local approximation to the modeled function that applies in the neighborhood of the new query instance (input vector) encountered, and never construct an approximation designed to perform well over the entire input space. Therefore, even a very complex target function can be described by constructing it from a collection of much less complex local approximations.

IBL algorithms have several advantages: they are quite simple but robust learning algorithms, can tolerate noise and irrelevant attributes, and can represent both probabilistic and overlapping concepts and naturally exploit inter-attribute relationships (Aha et al., 1991). However, they also have disadvantages. One of them is that the computational cost of predicting or classifying new instances can be very high since all computations takes place when the new instances have to be classified or predicted rather than when the training examples are first encountered. The computational cost increases with the number of training data and dimension of the input data and has order of $|T| \times p$, where T is the training set and p is the dimension of the input vector. Another disadvantage is that since all attributes of the examples are considered when attempting to retrieve similar training examples from the stored database, the examples that are truly most similar may well be a large distance apart if the target variable only depends on only a few of the many input variables.

The most common IBL methods used in numeric prediction are k-nearest neighbour method and locally weighted regression. The latter which was used in this study is described briefly in subsection 3.7.1. For the detailed description of IBL methods, the readers are referred to Aha et al. (1991). Application of these methods in rainfall-runoff modelling was recently reported in Solomatine et al. (2006, 2008).

3.7.1 Locally weighted regression

Locally weighted regression (LWR) is a memory-based method that performs a regression around a point x_q of interest (often called a query point) using only training data that are "local" to that point. The training examples are assigned weights according to their distance to the query instance, and regression equations are generated using the weighted data. The phrase "locally weighted regression" is called *local* because the function is approximated based on data near query point, *weighted* because the contribution of each training example is weighted by its distance from the query point.

The target function f in the neighbourhood surrounding the query point $\mathbf{x}_q$ can be approximated using a linear function, a quadratic function, neural network, etc.

A number of distance-based weighting schemes can be used in LWR (Scott, 1992). A common choice is to compute the weight w_i of each instance $\mathbf{x}_i$ according to the inverse of their Euclidean distance $d(\mathbf{x}_i, \mathbf{x}_q)$ from the query instance $\mathbf{x}_q$ as given by

$$w_i = K(d(\mathbf{x}_q, \mathbf{x}_i)) = (d(\mathbf{x}_q, \mathbf{x}_i))^{-1} \tag{3.8}$$

where $K(.)$ is typically referred to as the kernel function. The Euclidean distance $d(\mathbf{x}_i, \mathbf{x}_q)$ is define as

$$d(\mathbf{x}_i, \mathbf{x}_q) = \sqrt{\sum_{r=1}^{p}(a_r(\mathbf{x}_i) - a_r(\mathbf{x}_q))^2} \tag{3.9}$$

where an arbitrary instance $\mathbf{x}$ can be described by the feature vector $< a_1(\mathbf{x}), ..., a_p(\mathbf{x}) >$ and $a_r(\mathbf{x}_i)$ denotes the value of the rth attribute of the instance $\mathbf{x}_i$.

Alternatively, instead of weighting the data directly, the model errors for each instance used in the regression equation are weighted to form the total error criterion $C(q)$ to be minimized:

$$C(q) = \sum_{i=1}^{|T|} L\left(f(\mathbf{x}_i, \beta), y_i\right) K(d(\mathbf{x}_i, \mathbf{x}_q)) \tag{3.10}$$

where $f(\mathbf{x}_i, \beta)$ is the regression model; $L(.)$ is loss or error function (typically the sum of squared differences $(y_i - \hat{y}_i)^2$ between the target y_i and its estimated $\hat{y}_i$ values), β is a vector of parameters or coefficient of the model to be identified. Note that total error criterion is summed over the entire set T of the training examples. However there are other forms of the error criteria such as loss function over just the k nearest neighbours (Mitchell, 1997). In locally weighted regression, the function f is a linear function of the form:

$$f(\mathbf{x}_i, \beta) = \beta_0 + \beta_1 a_1(\mathbf{x}_i) + ... + \beta_p a_p(\mathbf{x}_i) \tag{3.11}$$

Cleveland and Loader (1994) among other addressed the issue of choosing weighting (kernel) functions: it should be maximum at zero distance, and the function should decay smoothly as the distance increases. Discontinuities in the weighting functions lead to discontinuities in the predictions, since the training points cross the discontinuity as the query changes. Yet another possibility to improve the accuracy of LWR is to use the so-called smoothing, or bandwidth parameter that scales the distance function by dividing it by this parameter. One way to choose it is to set it to the distance to the kth nearest training instance, so that its value becomes smaller as the volume of training data increases. Generally, an appropriate smoothing parameter can be found using cross-validation.

3.8   Clustering methods

The previous sections 3.5-3.7 present supervise learning methods in where the training data consist of pairs of input vectors, and target outputs. In contrast, clustering is an unsupervised learning method i.e. the target output is not known. In other words, training data in clustering consists of only unlabeled input vectors. Without knowing the target outputs of the data set, the best that can be done is to explore the data, to find natural groups within the data, and to derive some important features or characteristics of the groups.

The objective of cluster analysis is the attribution of objects to different groups, or more precisely, the partitioning of data set into subsets (clusters), so that the data in each subset share some common trait – often proximity according to some defined similarly measure. There are basically three types of clustering algorithms - exclusive, overlapping, and hierarchical.

Algorithms of the first type group the data in such a way that each data point belongs to a definite cluster and it could not be included in another cluster. The example of this type of clustering is *K-means* clustering.

Algorithms of the second type, result in the overlapping clusters and typically use fuzzy sets (Zadeh, 1965) to cluster data, so that each data point belongs to several clusters with some degree so called fuzzy membership in the range [0, 1]. The most known method of fuzzy clustering is the *fuzzy C-means* (Bezdek, 1981). There are also non-fuzzy clustering methods leading to overlapping clusters, such as those based on the mixture of Gaussians.

Hierarchical clustering finds successive clusters using previously established clusters. Such algorithm can be agglomerative ("bottom-up") or divisive ("top-down"). An agglomerative algorithm begins with each element as a separate cluster and merges them into successively larger clusters. A divisive algorithm begins with the whole set and proceed to divide it into successively smaller clusters.

Among all these algorithms the most widely used are K-means and fuzzy C-means methods and they are briefly described in the following subsections.

3.8.1   K-means clustering

K-means clustering algorithm aims at partitioning a set of given data matrix $X = \{x_1, ..., x_n\}$, where each data point x_j is a P-dimensional vector, into c clusters C_i by minimizing an objective function J:

$$J = \sum_{i=1}^{c} \sum_{j \in C_i} \left\| \mathbf{x}_j - \mathbf{v}_i \right\|^2 \tag{3.12}$$

where $\| \, . \, \|$ is the Euclidean norm, $\mathbf{v}_i$ is the coordinate (in P-dimension) of centre of ith cluster. The idea is to find the best partitions of N samples by c clusters C_i such that the total distance between the clustered samples and their respective centres (i.e. total variance) is minimised.

K-means algorithm is very simple. It starts by selecting c points as initial cluster centres either at random or using some heuristic data. Instances are assigned to their closest centre according to the Euclidean distance function. Next the centroid or mean of all instances in each cluster is calculated. These centroids are taken to be the new centres for their respective clusters. Finally the whole process is repeated with the new cluster centres until the value of J becomes stable and consequently the centres stop to move.

3.8.2 Fuzzy C-means clustering

The most known method of fuzzy clustering is the fuzzy C-means (FCM), initially proposed by Dunn (1973) and generalized by Bezdek (1981). FCM algorithm attempts to partition a set of given data $X = \{\mathbf{x}_1, ..., \mathbf{x}_n\}$ where each data point $\mathbf{x}_k$ is a vector in R^p. The partitioned space M_{fc} takes the form of

$$M_{fc} = \left\{ \begin{array}{l} P_{i,j} \in R^{c \times n} : \mu_{i,j} \in [0,1] \;\; \forall i, j \\ \sum_{i=1}^{c} \mu_{i,j} = 1 \;\; \forall j, \;\; 0 < \sum_{j=1}^{n} \mu_{i,j} < n \;\; \forall i \end{array} \right\} \tag{3.13}$$

where c is the number of clusters, n is the number of data points, P is the partition matrix that contains the membership values $\mu_{i,j}$ of each data j for each cluster i. The aim of the FCM algorithm is to find an optimal fuzzy C-means partition that minimizes the following objective function:

$$J_m(P,V) = \sum_{i=1}^{c} \sum_{j=1}^{n} \mu_{i,j}^{m} \left\| \mathbf{x}_j - \mathbf{v}_i \right\|^2 \tag{3.14}$$

where $V = (\mathbf{v}_1, ..., \mathbf{v}_c)$ is a matrix of unknown cluster centres (prototypes) $\mathbf{v}_i \in \Re^p$, and the weighting exponent m in $[1, \infty]$ is a constant that influences the membership values. With fuzzy C-means, the centroid of a cluster is the mean of all points, weighted by their degree of belonging to the cluster:

$$\mathbf{v}_i = \sum_{j=1}^{n} \mu_{i,j}^m \mathbf{x}_j \bigg/ \sum_{j=1}^{n} \mu_{i,j}^m \qquad (3.15)$$

The degree of belonging i.e. the membership degree $\mu_{i,j}$ is related to the inverse of the distance between the data point $\mathbf{x}_j$ and the cluster centre $\mathbf{v}_i$:

$$\mu_{i,j} = \left[\sum_{k=1}^{c} \left(\frac{\left\| \mathbf{x}_j - \mathbf{v}_i \right\|^2}{\left\| \mathbf{x}_j - \mathbf{v}_k \right\|^2} \right)^{2/(m-1)} \right]^{-1} \qquad (3.16)$$

Here the distance is fuzzyfied with a real parameter $m > 1$ and normalised so that their sum is 1. For m equal to 2, this is equivalent to normalising the coefficient linearly to make their sum 1. When m is close to 1, then cluster centre closest to the point is given much more weight than the others, and the algorithm is similar to K-means.

Working of the fuzzy C-means algorithm is very similar to that of K-means algorithm, and partitioning is carried out through an iterative optimisation of the objective function of equation (3.15), with the update of membership $\mu_{i,j}$ and the cluster centres $\mathbf{v}_i$. The algorithm is composed of the following steps:

1. Input:

   - Data $X = \{\mathbf{x}_1, ..., \mathbf{x}_n\}$, number of clusters c, fuzzification parameter m (with $m > 1$) and small positive constant ε for stopping criteria

2. Output:

   - Cluster centre $\mathbf{v}_i$ and matrix of membership values P.

3. Initialise:

   - Iteration t = 1

   - Generate randomly initial cluster centres $\mathbf{v}_i^{(t)}$ (i=1, ..., c)

   - Compute matrix $P^{(t)}$ of membership values $\mu_{i,j}$ using equation (3.16) $\forall j$

4. Iterate until $| P^{(t+1)} - P^{(t)} | > \varepsilon$

   - $t=t+1$.

   - Given the membership value $\mu_{i,j}$ update the cluster centres $\mathbf{v}_i^{(t+1)}$ using equation (3.15).

   - Update the membership value $\mu_{i,j}^{(t+1)}$ with newly computed cluster centres using equation (3.16).

3.8.3   Validity measures

Since we usually have little information about the data structure in advance, a crucial step in the clustering is determining the optimal number of clusters. This has turned out to be a rather difficult problem, as it depends on the structure of the cluster. Without the prior information, a common method is the comparison of partitions resulting from different number of clusters. Different validity measures are reported in the literature to find the optimal number of clusters such as the partition coefficient, PC (Bezdek, 1981), partition index, SC, separation index, S (Bensaid et al., 1996), Xie-Beni's index, XB (Xie and Beni, 1991). Since none of these measures is reliable and perfect, the optimal number of clusters obtained through the validity measures has to be checked through the final results of application where the clustering technique was applied (Shrestha and Solomatine, 2008).

Partition index (SC) is the ratio of sum of compactness and separation of the clusters. This is given by:

$$SC(c) = \sum_{i=1}^{c} \frac{\sum_{j=1}^{n} \mu_{i,j}^{m} \left\| x_j - v_i \right\|^2}{n_i \sum_{k=1}^{c} \left\| v_k - v_i \right\|^2} \tag{3.17}$$

Lower value of SC indicates a better partition. On the contrary of partition index, separation index (S) uses a minimum distance separation for partition validity and is given by

$$S(c) = \frac{\sum_{i=1}^{c} \sum_{j=1}^{n} \mu_{i,j}^{2} \left\| x_j - v_i \right\|^2}{n \min_{k \neq i} \left(\left\| v_k - v_i \right\|^2 \right)} \tag{3.18}$$

Xie and Beni's index (XB) aims to quantify the ratio of the total variation within clusters and the separation of clusters and is given by:

$$XB(c) = \frac{\sum_{i=1}^{c} \sum_{j=1}^{n} \mu_{i,j}^{m} \left\| x_j - v_i \right\|^2}{n \min_{i \neq j} \left(\left\| x_j - v_i \right\|^2 \right)} \tag{3.19}$$

The optimal number of clusters should minimize the value of XB index.

Beside the number of clusters, an important step in any clustering is to select a suitable distance measure, which will determine how the *similarity* of any two data points is calculated. This will influence the shape of the clusters, as some points may be close to one another according to one distance and further away according to another. The common distance function is the Euclidean distance or the squared Euclidean distance.

3.9 Selection of input variables

In machine learning techniques, the selection of appropriate model inputs is extremely important as they contain the important information about the complex (linear or non-linear) relationship with the model outputs. The objective of the input variable selection is three-fold (Guyon and Elisseeff, 2003): improving the prediction performance of the predictor (here machine learning models), providing faster and more cost effective predictors, and providing a better understanding of the underlying process that generated the data. Any input variable selection method should follow the Occam's Razor principle to build as simple a model as possible that still achieves some acceptable level of the performance on the training data. Unlike the physically based models, the sets of input variables that influence the system output are not known a priori. Therefore the selection of the appropriate model inputs that will allow machine learning techniques to map to the desired output vector successfully is not a trivial task. Usually, not all of the potential input variables will be equally informative since some may be correlated, noisy or have no significant relationship with the output variables being modeled. Irrelevant and redundant variables also may confuse the learning machines by helping to obscure the distribution of the small set of truly relevant input variables (Koller and Sahami, 1996).

Bowden et al. (2005) reviewed some of the input variables selection methods used in water resources ANN applications. These methods are also applicable to other supervised machine learning methods described in this chapter. These methods can be broadly classified into the three groups:

A priori or domain knowledge: The importance of using such knowledge in selecting appropriate input variables is obvious. A good level of understanding of the system (e.g. hydrological system for rainfall-runoff modelling) being modelled is very important in order to select the appropriate model inputs (ASCE, 2000). This would help in avoiding loss of information that may result if key input variables are omitted, and also prevent inclusion of spurious input variables that tend to confuse the training process (Sudheer et al., 2002). Even in the other methods described below, some form of domain knowledge is still used. Although domain knowledge is often used in machine learning methods to select the input variables, it is dependent on an expert's knowledge and hence, is very subjective and case dependent.

Analytical and visual techniques: When the relationship to be modelled is not well understood, then analytical techniques such as correlation and mutual information analysis are employed. These techniques can be used to determine the strength of the relationship between the input time series variables and the output time series at various lags. While correlation analysis is used to find the linear relationship between the variables, mutual information analysis is used to determine linear or non-linear the dependencies. Correlation analysis consists of computing cross correlation between the input vector x_i and the output variable y. In case of time series variable, it is often required to compute a) correlation of lagged vector of x_i with the output y; and b) autocorrelation of the output vector y. The former measures the dependency of the output variable to the previous values of the input variables. The latter provides the information on the dependency of output variable on its past values.

Correlation coefficient (CoC) between input vector x_t and output vector y_t, t=1, ..., n is given by:

$$CoC = \frac{\sum_{i=1}^{n}(x_i-\bar{x})(y_i-\bar{y})}{\sqrt{\sum_{i=1}^{n}(x_i-\bar{x})^2}\sqrt{\sum_{i=1}^{n}(y_i-\bar{y})^2}} \tag{3.20}$$

where $\bar{x}$ is the mean of x. The maximum value of CoC is 1 for complete positive correlation and the minimum value of it is -1 for complete negative correlation. A value of CoC close to zero indicates that the variables are uncorrelated.

Mutual information which is based on Shannon's entropy (Shannon, 1948) is used to investigate linear and non-linear dependencies and lag effects (in time series data) between the variables. The mutual information is measure of information available from one set of data having knowledge of another set of data. The average mutual information AMI between two variables X and Y is given by

$$AMI = \sum_{i,j} P_{XY}(x_i,y_j)\log_2\left[\frac{P_{XY}(x_i,y_j)}{P_X(x_i)P_Y(y_j)}\right] \tag{3.21}$$

where $P_X(x)$ and $P_Y(y)$ are the marginal probability density functions of X and Y, respectively, and $P_{XY}(x,y)$ is the joint probability density functions of X and Y. If there is no dependence between X and Y, then by definition the joint probability density $P_{XY}(x,y)$ would be equal to the product of the marginal densities $(P_X(x) P_Y(y))$. In this case, AMI would be zero (the ratio of the joint and marginal densities in equation (3.21) being one, giving the logarithm a value of zero). A high value of AMI would indicate a strong dependence between two variables. The key to accurate estimate of the AMI is the accurate estimation of the marginal and joint probabilities density in equation (3.21) from a finite set of examples. The most widely used approach is estimation of the probability densities by histogram with the fixed bin width. More stable, efficient and robust probability density estimator is based on the use of kernel density estimation techniques (Sharma, 2000).

Heuristic approaches: Several machine learning models are trained using different subsets of inputs, and the inputs which give the best performance are considered as the plausible inputs for the model. Other commonly used heuristic approaches are the stepwise forward and backward selection. The former approach starts by finding the best single inputs and considering it for the final model. In each subsequent step, given a set of selected inputs, the input that improves the model's performance is added to the final model. The latter approach starts with a set of all inputs, and subsequently eliminates the input that reduces the performance the least. The main disadvantage of the heuristic approaches being trial-and-error is that they are computationally intensive (Bowden et al., 2005). All the mentioned methods are sometimes called "model-based" since they require running the model.

In the light of non-linear and complex relationships between the variables, the input selection from the finite data sets based on only one method does not provide guarantee that they will find the globally optimal sets of inputs. Intuitively, the preferred approach for determining appropriate inputs and lags of inputs involves a combination of (some of the) above methods, and it was used in this thesis.

Chapter 4
Machine Learning in Prediction of Parameter Uncertainty: MLUE Method

This chapter presents a novel approach to model and predict parametric uncertainty of rainfall-runoff model. The most widely used method to analyse such uncertainty (Monte Carlo simulation) involves multiple model runs. The computational load may prohibit the use of such method for real-time predictions of uncertainty valid for particular combinations of the input and state variables (hydrometeorological conditions). Machine learning techniques allow for the building of surrogate models for the results of such simulations; these models are fast and can be easily used in operation for real-time predictions of parameter uncertainty. The proposed method is referred to as MLUE (Machine Learning in parameter Uncertainty estimation).

4.1   Introduction

As mentioned in Chapter 1, hydrological models, in particular rainfall-runoff models, are simplified representations of the reality and aggregate the complex, spatially and temporally distributed physical processes through relatively simple mathematical equations with parameters. The parameters of the rainfall-runoff models can be estimated in two ways (Johnston and Pilgrim, 1976). First, their values are estimated from available knowledge or measurements of the physical process, provided the model parameters realistically represent the measurable physical process. In the second approach, parameter values are estimated using input and output measurements, where the parameters do not represent directly measurable entities or when it is too costly to measure them in the field. Conceptual rainfall-runoff models usually contain several such parameters, which cannot be directly measured. The process of estimating the parameters is called model *calibration* and it has received increased scientific interest in the last two decades (e.g., Duan et al., 1992; Yapo et al., 1996; Solomatine et al., 1999; Madsen, 2000). The calibration technique adjusts the parameters of the model in such a way that the behaviour of the model approximates as closely and consistently as possible the observed responses of the hydrological system over some historical period of time. Traditional calibration procedures, which involve a *manual* adjustment of the parameter values, are labour-intensive, and their success is strongly dependent on the experience of the modeller. Because of the time consuming nature of manual calibration, there has been a great deal of research into the development of automatic calibration methods. Automatic methods for model calibration seek to take advantage of the speed and power of computers, while being relatively objective and easier to implement than manual methods (Vrugt et al., 2005).

While considerable attention has been given to the development of automatic calibration methods which aim to find successfully the single best set of values for the parameter vector, much less attention has been given to a realistic assessment of parameter uncertainty in hydrological models. Since the model never perfectly represents the system, a single set of the parameter values obtained from the calibration may not represent a true parameter set, particularly as the data used for calibration is subjected to measurement errors. Hence, the model prediction based on the single parameter set may have considerable uncertainty. Rather, it is required to have a probability distribution of the parameters which can describe the uncertainty associated with the parameter estimation, and then such parameter uncertainty can be propagated through the model to deduce a probability distribution of the model output.

4.2   Monte Carlo techniques for parameter uncertainty analysis

The uncertainty analysis of rainfall-runoff models has been received special attention in recent years. Several uncertainty analysis methods have been developed to propagate the uncertainty through these models and to derive meaningful uncertainty bounds of the model simulations. We presented these methods in Chapter 2. In this chapter we discuss the Monte Carlo based uncertainty analysis methods, which have been mostly used for parameter uncertainty analysis.

Monte Carlo (MC) simulation techniques have been applied for uncertainty analysis very successfully in the last decades in hydrological sciences, and have lead the quantification of the model output uncertainty resulting from uncertain model parameters, input data or model structure. The technique involves random sampling from the distributions of uncertain inputs and the model is run successively until a desired statistically significant distribution of outputs is obtained. The main advantage of the MC simulation based uncertainty analysis is its general applicability; however, methods of this type require a large number of samples (or model runs), and their applicability is sometimes limited to simple models. In the case of computationally intensive models, the time and resources required by these methods could be prohibitively expensive.

A version of the MC simulation method was introduced under the term "generalised likelihood uncertainty estimation" (GLUE) by Beven and Binley (1992). GLUE is one of the popular methods for analysing parameter uncertainty in hydrological modelling and has been widely used over the past ten years to analyse and estimate predictive uncertainty, particularly in hydrological applications (see, e.g., Freer et al., 1996; Beven and Freer, 2001; Montanari, 2005). Users of GLUE are attracted by its simple understandable ideas, relative ease of implementation and use, and its ability to handle different error structures and models without major modifications to the method itself. Despite its popularity, there are theoretical and practical issues related with the GLUE method reported in the literature. For instance, Mantovan and Todini (2006) argue that GLUE is inconsistent with the Bayesian inference processes such that it leads to an overestimation of uncertainty, both for the parameter uncertainty estimation and the predictive uncertainty estimation. For issues regarding inconsistency and other criteria, readers are referred to the citation above and the subsequent discussions in the Journal of Hydrology in 2007 and 2008. A practical problem with the GLUE method is that, for models with a large number of parameters, the sample size from the respective parameter distributions must be very large in order to achieve a reliable estimate of model uncertainties (Kuczera and Parent, 1998). In order to reduce the number of samples (i.e., model runs) a hybrid genetic algorithm and artificial neural network has been applied by Khu and Werner (2003). Similarly, Blasone et al. (2008) used adaptive Markov chain Monte Carlo sampling within the GLUE methodology to improve the sampling of the high probability density region of the parameter space. Another practical issue is that in many cases the percentage of observations falling within the prediction limits provided by GLUE is much lower than the given confidence level used to produce these prediction limits (see, e.g., Montanari, 2005). Xiong and O'Connor (2008) modified the GLUE method to improve the efficiency of the GLUE prediction limits in enveloping the observed discharge.

The MC based method for uncertainty analysis of the outputs of hydrological models is very flexible, conceptually simple and straightforward, but becomes impractical in real time applications when there is little time to perform the uncertainty analysis because of the large number of model runs required. For such situations alternative approximation methods have been developed, referred to as moment propagation techniques, which are able to calculate directly the first and second moments without the application of an MC simulation (see, e.g., Rosenblueth, 1981; Harr, 1989; Protopapas and Bras, 1990; Melching, 1995; Kunstmann et al., 2002). A number of research studies have been conducted to reduce the number of MC simulation runs effectively, for instance, Latin hypercube sampling (see, e.g., McKay et al., 1979). Recently, new methods have been developed to improve the efficiency of the MC based uncertainty analysis method. Examples of such methods are the delayed rejection adaptive Metropolis method (Haario et al., 2006), and the differential evolution adaptive Metropolis method, DREAM (Vrugt et al., 2008b). In these Metropolis uncertainty analysis methods, the comparison of the statistics of multiple sample chains in parallel would provide a formal solution to assess how many model runs are required to reach convergence and obtain stable statistics of the model output and parameters. However, it is well recognized that traditional MC based simulation still lacks a well-established convergence criterion to terminate the simulations at a desired level of accuracy (e.g., Ballio and Guadagnini, 2004).

4.3 Problems attached to Monte Carlo based uncertainty analysis methods

The MC based methods for uncertainty analysis of the outputs of the process models are very flexible, robust, conceptually simple and straightforward, yet there are a few practical problems associated with the implementation of the methods:

The random sampling from a probability distribution in standard MC simulation is not very efficient. The convergence of the MC simulation is quite slow with the order $O(1/\sqrt{s})$ and a larger number of model runs is needed to establish a reliable estimate of uncertainties, where s is the number of simulations. This implies that, in order to increase accuracy of estimation by a factor of 2, one needs to increase the number of MC simulations by a factor of 4.

The number of simulations increases exponentially with the dimension of the parameter vector $O(n^p)$, where p is the dimension of parameter vector, and n is the number of samples required for each parameter. The total number of parameter combinations required to cover the full space of uncertainty quickly becomes prohibitive as more parameters are considered. Failure to maintain an adequate sampling density may result in undersampling probable regions of the parameter space (Kuczera and Parent, 1998). For example, if we wish to sample, on average, 5 points in each dimension in a 10 dimensional parameter space, we would need 5^{10} MC simulations. Considering a 0.001 second CPU time for a single run of the model for one time step, the total CPU time for 5^{10} MC simulations would be 1 day, which is impracticable in many cases (e.g., real time flood forecasting system).

4.4 Machine learning emulator in uncertainty analysis

As mentioned above, a number of new methods have been developed to improve the efficiency of MC based uncertainty analysis methods and yet these methods still require a considerable number of model runs in both offline and operational mode to produce reliable and meaningful uncertainty estimation. In this thesis we use a different approach to uncertainty analysis. Our approach is to use machine learning techniques to emulate the MC simulation results. The proposed method is referred to as the MLUE (**M**achine **L**earning in parameter **U**ncertainty **E**stimation).

The proposed technique to emulate the complex model by a simple model is an example of surrogate modelling, or meta-modelling - an approach widely used when running the complex model is computationally expensive. For example, O'Hagan (2006) used the Gaussian process emulator to emulate a complex simulation model. Li et al. (2006) introduced a new approach to meta modelling whereby a sequential technique is used to construct and simultaneously update mutually dependent meta-models for multiresponse, high-fidelity deterministic simulations. Young and Ratto (2009) proposed a dynamic emulation model to emulate a complex high order model by a low order data based mechanistic model. The novelty of our method is in the following:

1. The MLUE method explicitly builds an emulator for the MC uncertainty results while other methods build an emulator for a single simulation model.

2. The MLUE emulator is based on machine learning techniques, while other techniques are Bayesian (e.g., O'Hagan, 2006), or use nonlinear differential equations (e.g., the data based mechanistic model of Young (1998)).

4.5   Methodology

The idea of using statistical and, in general, machine learning models, to improve model accuracy is not new. Typically, information about model errors is used to train machine learning error correctors (Abebe and Price, 2003) or to build more sophisticated machine learning models of model uncertainty (Shrestha and Solomatine, 2006a, 2008). In this chapter, we extend this idea towards building a model (referred to as U) encapsulating the information about the realizations of the process (e.g., hydrological) model M output generated by MC simulations. Instead of predicting a single value of the model error, as done in the most error correction procedures, we aim to predict the distribution of the output of M generated by MC based simulations. Thus, the method allows one to predict the uncertainty bounds of the model M prediction without re-running the MC simulations when new input data is observed and fed into M, while the MC based uncertainty analysis methods require a fresh set of MC runs for each analysis. For instance, GLUE will typically require a fresh set of MC runs from the behavioural models to produce the prediction intervals for the model output for each time step with the new data input.

The basic idea here is to estimate the uncertainty of the model M output assuming that uncertainty at a particular time step depends on the corresponding forcing input data and the model states (e.g., rainfall, antecedent rainfall, soil moisture etc.). We also assume that the uncertainty associated with the prediction of the hydrological variables such as runoff in similar hydrological conditions is also similar. By "hydrological conditions" we mean here the combination of the state of the input data and the state variables, which are driving the generation of the runoff in the catchment. For example, one can see that, compared to the low flows, it is more difficult to predict extreme events such as peak flows. Consequently, uncertainty of the prediction of the peak flows is higher compared to those for low flows. In other words, we assume that there is a dependency between the model uncertainty statistics and the input data of the process model, and this dependency might be sufficient to build a machine learning model that can predict the model uncertainty when the input data vector is presented. The flow chart of the MLUE methodology is presented in Figure 4.1.

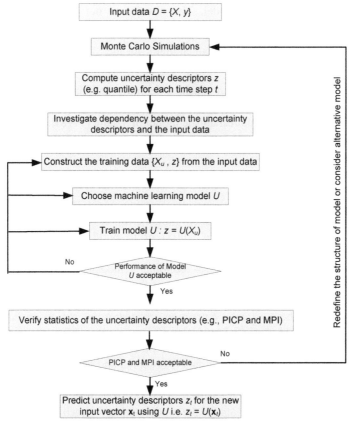

Figure 4.1. Schematic diagram of using machine learning method to estimate uncertainty generated by MC simulations.

4.5.1 Monte Carlo simulations

Consider a deterministic model M of a real world system predicting a system output variable y given the input data X, the initial condition of the state variables s_0 and the vector of the parameter θ. The model M could be physically based, conceptual, or even data- driven. The model M is also referred to as "primary model" in order to distinguish from uncertainty model which will be described later. For the sake of simplicity, the model M here is referred to as a conceptual rainfall-runoff model. The system response can be represented by the following equation:

$$y = M(\mathbf{x}, s, \theta) + \varepsilon = \hat{y} + \varepsilon \tag{4.1}$$

where ε is the vector of the model errors between the vector of the observed response y and the corresponding model response $\hat{y}$. Note that the state variable s, which appears in the equation (4.1), will be computed by running the model M given the initial condition of the state variables s_0. Before running the model M, the components of the model i.e. input data vector $\mathbf{x}$, initial conditions s_0, parameter θ, and the model structure itself, have to be specified, while the output or model response $\hat{y}$ and the state variable s are computed by running the model. These components may be uncertain in various ways to various degrees; the consequences of these uncertainties will be propagated into the model states and the outputs.

The MC simulation is performed by running the model multiple times by sampling either the input data $\mathbf{x}$ or the parameters vectors θ or even the structure of the model, or a combination of them. Thus, mathematically it is equivalent to the following:

$$\tilde{\hat{y}} = \tilde{M}(\tilde{\mathbf{x}}, \tilde{\theta}) \tag{4.2}$$

where $\tilde{M}$ is the possible candidate of the model structure, $\tilde{\mathbf{x}}$ is the sampled input data from the given pdf, $\tilde{\theta}$ is the parameter sampled with the given pdf from the feasible domain of the parameter space. For the sake of simplicity we assume that model structure and input data are correct, then equation (4.2) can be rewritten as:

$$\hat{y}_{t,i} = M(\mathbf{x}, \theta_i); \quad t=1, ..., n; \quad i=1, ..., s \tag{4.3}$$

where θ_i is the set of parameters sampled for the ith run of MC simulation, $\hat{y}_{t,i}$ is the model output of the tth time step for the ith run, n is the number of time steps and s is the number of simulations.

4.5.2 Characterisation of uncertainty

Having enough realizations of the MC simulations, the desired statistical properties such as moments and quantiles of the model output for each time step t are estimated from the realizations $\hat{y}_{t,i}$. One way to judge the uncertainty of the model output is to use the second moment (variance) of the model error: a large variance of the model error typically indicates that the model prediction is uncertain. In most cases, however, the variance does not sufficiently describe the uncertainty. Furthermore if the model errors are not Gaussian, which is often the case (see, e.g., Xu, 2001; Shrestha and Solomatine, 2008), then the error variance cannot be used to derive the predictive uncertainty. In this situation more informative quantities such the cumulative distribution function (cdf) or some of its quantiles or prediction intervals are used. We denote such a quantity by z which characterises the uncertainty of the model prediction. The following uncertainty descriptors can be foreseen:

1. The prediction variance $\sigma_t^2(\hat{y}_t)$

$$\sigma_t^2(\hat{y}_t) = \frac{1}{s-1} \sum_{i=1}^{s} (\hat{y}_{t,i} - \bar{\hat{y}}_{t,i})^2 \tag{4.4}$$

where $\bar{\hat{y}}_{t,i}$ is the mean of MC realizations at the time step t.

2. The prediction quantile $\hat{Q}_t(p)$ corresponding to the pth [0, 1] quantile

$$P(\hat{y}_t < \hat{Q}_t(p)) = \sum_{i=1}^{s} w_i \left| \hat{y}_{t,i} < \hat{Q}_t(p) \right| \tag{4.5}$$

where w_i is the weight given to the model output at simulation i, $\hat{y}_t$ is the realization vector at time step t, $\hat{y}_{t,i}$ is the value of model outputs at the time t simulated by the model $M(\mathbf{x}, \theta_i)$.

3. The transferred prediction quantile $\Delta\hat{Q}_t(p)$ corresponding to the pth [0, 1] quantile

$$\Delta\hat{Q}_t(p) = \hat{Q}_t(p) - \overline{y}_t \tag{4.6}$$

where $\overline{y}_t$ is the output of the calibrated (optimal) model. Note that the quantiles $\Delta\hat{Q}_t(p)$ obtained in this way are conditional on the model structure, inputs and the likelihood weight vector w_i. The essence of this transferred prediction quantile will be apparent in the next section.

4. The prediction intervals $[PI_t^L(\alpha) \quad PI_t^U(\alpha)]$ derived from the transferred prediction quantile for given confidence level of 1-α (0<α<1)

$$PI_t^L(\alpha) = \Delta\hat{Q}_t(\alpha/2), \quad PI_t^U(\alpha) = \Delta\hat{Q}_t((1-\alpha)/2) \tag{4.7}$$

where $PI_t^L(\alpha)$ and $PI_t^U(\alpha)$ are the distance between the model output to the lower and upper prediction limits respectively and referred to the lower and upper prediction intervals corresponding to the 1-α confidence level (although, formally, these are not intervals but distances).

4.5.3 Predictive model for emulating MC simulations

Once the desired uncertainty descriptors are computed from the realizations of MC simulations, machine learning models are used to emulate the MC simulations by estimating the uncertainty descriptors. The machine learning models are used to map the input data and state variables (if any) to the uncertainty descriptors of the model output that is generated by the MC simulations. Experience suggests that the model residuals (errors) may show a non-stationary bias, heteroscedasticity, non-stationary skewness and autocorrelation over one or more time steps (Beven and Freer, 2001). This characteristic of the model output distribution motivates us to build the machine learning models to approximate not only the mean and variance, but also the other uncertainty descriptors.

The machine learning model U to learn the functional relationship between the input data $\mathbf{x}$ and the uncertainty descriptor z takes the form:

$$z_t = U(\mathbf{x}_t) + \xi_t \tag{4.8}$$

where ξ_t is the residual (error) between the target uncertainty descriptor z_t and the predicted uncertainty descriptor by the machine learning model. The input data vector $\mathbf{x}$ used to train the machine learning models is typically different from the input to the process based model M and is discussed in section 4.6. The input data $\mathbf{x}$ is constructed from the input variables of the process model, state variables, lagged variables of input and output and other relevant variables that could help to increase accuracy of the prediction. The residual ξ measures the accuracy and predictability of the machine learning model U. Model U, after being trained, encapsulates the underlying dynamics of the uncertainty descriptors of the MC simulations and maps the input to those descriptors. The model U can be of various types, from linear to non-linear regression models such as an ANN. The choice of model depends on the complexity of the problem to be handled and the availability of data. Once the model U is trained on the calibration data, it can be employed to estimate the uncertainty descriptors such as prediction intervals for the new input data vector that was not used in any of the model building process (see Figure 4.1).

If the uncertainty descriptor is the transferred prediction quantile (equation (4.6)), then the model U will take the form:

$$\Delta \hat{Q}_t(p) = U(\mathbf{x}_t) + \xi_t \tag{4.9}$$

If the uncertainty descriptor is the prediction interval derived from the transferred prediction quantile (equation (4.7)), then the model U will take the form:

$$PI_t^L(\alpha) = U_L(\mathbf{x}_t) + \xi_L \tag{4.10}$$
$$PI_t^U(\alpha) = U_U(\mathbf{x}_t) + \xi_U$$

Since the transferred prediction quantile is derived from the contemporary value of the model simulations (equation (4.6)), then the predictive quantile of the model output is estimated by

$$\hat{Q}_t(p) = U(\mathbf{x}_t) + \bar{y}_t \tag{4.11}$$

Similarly upper and lower prediction limits of the model output is given by

$$PL_t^L(\alpha) = U_L(\mathbf{x}_t) + \bar{y}_t \tag{4.12}$$
$$PL_t^U(\alpha) = U_U(\mathbf{x}_t) + \bar{y}_t$$

where U_L and U_U are the machine learning models for the lower and upper prediction intervals, respectively. It is worthwhile to mention that for the uncertainty descriptors of equation (4.6) and equation (4.7), it is assumed that there is an optimal model.

4.6   Selection of input variables

As mentioned in section 3.9, the selection of appropriate model inputs is extremely important as they contain important information about the complex (linear or non-linear) relationship with the model outputs. Therefore the success of the MLUE method depends on the appropriate selection of the input variables to use in machine learning model U. The required input variables can vary depending on the type of the process model and the inputs used in the process model, among others. In most cases, a combination of the domain knowledge and the analytical analysis of the causal relationship may be used to select relevant variables to use as the input to the machine learning model. The input variables to the machine learning model that can be considered are termed *plausible data* and include:

- Input variables to the process or primary model;

- State variables;

- Observed outputs of the process model;

- Outputs of the process model;

- Time derivatives of the input data and state variables of the process model;

- Lagged variables of input, state and observed output of the process model; and

- Other data from the physical system that may be relevant to the pdf of the model errors.

In most practical cases, the input data set **x** can be constructed from the plausible data set according to the methods discussed in section 3.9. Since the natures of the models M and U are very different, an analytical technique such as linear correlation or average mutual information between the quantiles of the model error and the plausible data is needed to choose the relevant input variables. As note in the above list of the plausible data, the input variable might also consist of the lagged variables of input, state and observed output of the process model. Based on the domain knowledge and the analytical analysis of the causal relationship, several structures of input data can be tested to select the optimal input data structure.

For example, if the model M is a conceptual hydrological model, it would typically use rainfall (R_t) and evapotranspiration (E_t) as input variables to simulate the output variable runoff (Q_t). However, the uncertainty model U, whose aim is to uncertainty of the simulated runoff, may be trained with the possible combination of rainfall and evapotranspiration (or effective rainfall), their past (lagged) values, the lagged values of runoff, and, possibly, their combinations. Let us recall equation (4.8) of uncertainty model U and extend it:

$$z_t = U(R_t', EP_t', Q_{t-1}', S_t', ...) + \xi_t \tag{4.13}$$

where,

$R_t' = R_t, R_{t-1}, ..., R_{t-\text{amax}}$ is the lagged inputs of the rainfall

$EP_t' = EP_t, EP_{t-1}, \ldots, EP_{t-\bar{m}\max}$ is the lagged inputs of the potential evapotranspiration

$Q_{t-1}' = Q_{t-1}, Q_{t-2}, \ldots, Q_{t-\bar{m}\max}$ is the lagged inputs of the runoff

$S_t' = S_t, S_{t-1}, \ldots, S_{t-\bar{m}\max}$ is the lagged inputs of the state variable (e.g., soil moisture etc)

We follow above conventions throughout this thesis in defining lagged inputs of the variables. The difficulty here is to select appropriate lags $\bar{m}$max for each input variables beyond which, values of the input time series have no significant effect on the output time series (in our case uncertainty descriptors). A subset of inputs for the model U is selected based on methods discussed in section 3.9. It is noteworthy to mention that inputs to the model U should not include those variables which are not available or cannot be measured at the time of prediction. Thus in above formulation, Q_t is not included, however lagged of Q_t can be used as one of the inputs as shown above.

4.7   Validation of methodology

The uncertainty model U can be validated in two ways: (i) by measuring its predictive capability; and (ii) by measuring the statistics of the uncertainty. The former approach measures the accuracy of uncertainty models in approximating the uncertainty descriptors of the realizations of the MC simulations. The latter approach measures the goodness of the uncertainty models as uncertainty estimators.

Two performance measures such as the coefficient of correlation (CoC) and the root mean squared error (RMSE) are used to measure the predictive capability of the uncertainty model and are given by:

$$\text{CoC} = \frac{\sum_{t=1}^{n}(z_t-\bar{z})(V(\mathbf{x}_t)-\bar{V}(\mathbf{x}_t))}{\sqrt{\sum_{t=1}^{n}(z_t-\bar{z})^2}\sqrt{\sum_{t=1}^{n}(V(\mathbf{x}_t)-\bar{V}(\mathbf{x}_t))^2}} \tag{4.14}$$

$$\text{RMSE} = \sqrt{\frac{1}{n}\sum_{t=1}^{n}(z_t - V(\mathbf{x}_t))^2} \tag{4.15}$$

where $\bar{z}$ and $\bar{V}(\mathbf{x}_t)$ are the mean of the uncertainty descriptors and the mean of the uncertainty descriptors predicted by the uncertainty model U, respectively. Beside these numerical measures, the graphical plots such as the scatter and time plot of the uncertainty descriptors obtained from the MC simulation and their predicted values are used to judge the performance of the uncertainty model U.

The goodness of the uncertainty models is evaluated based on uncertainty measures by using the so-called prediction interval coverage probability (PICP) and mean prediction interval (MPI) (Shrestha and Solomatine, 2006a). The PICP is the frequency of the observed outputs falling within the computed prediction intervals corresponding to the prescribed confidence level of 1-α (say 90%). Thus, PICP measures the efficiency to bracket the observed outputs within the uncertainty bounds against the specified α value. PICP is given by:

$$\text{PICP} = \frac{1}{n}\sum_{t=1}^{n} C$$

$$\text{with } C = \begin{cases} 1, & PL_t^L \leq y_t \leq PL_t^U \\ 0, & \text{otherwise} \end{cases}$$

(4.16)

where y_t is the observed model output at the time t. Theoretically, the value of PICP should be close to the prescribed degree of confidence 1-α. However, if the value of PICP obtained by MC simulation is not close to 1-α, then we cannot expect that the PICP value obtained by uncertainty model U will be close to α. If the model U is good enough then its PICP value should be close to that of the MC simulations.

MPI is the average width of the prediction intervals and gives an indication of how high the uncertainty is:

$$\text{MPI} = \frac{1}{n}\sum_{t=1}^{n}(PL_t^U - PL_t^L)$$

(4.17)

The larger the uncertainty, the larger the value of the MPI will be. In the ideal case if there is no uncertainty, then the value of the MPI will be zero. However, the MPI alone does not say too much; it will be used together with PICP to compare the performance of the uncertainty models. The best model will be one which gives the PICP close to 1-α with the lowest MPI. It is obvious that the PICP will be increased with the increase of the MPI.

Beside these uncertainty statistics, the plot of uncertainty bounds and the observed model output are investigated to judge the performance of the uncertainty model. Visual inspection of these plots can provide significant information about how effective the uncertainty model is in enclosing the observed outputs along the different input regimes (e.g., low, medium or high flows in hydrology).

4.8   Limitations of the method

This section discusses some issues concerning the limitations and possible extensions of the method. Since the machine learning technique is the core of the method, the problem of extrapolation, which is a well known problem of machine learning techniques, is also present. This means that the results are reliable only within the boundaries of the domain where the training data are given. In order to avoid the problem of extrapolation, an attempt should be made to ensure that the training data includes all possible combinations of the events including the extreme (such as extreme flood), and this is not always possible since the extremes tend to be rather rare events. Like the conventional approach of uncertainty analysis, the MLUE method also presupposes the existence of a reasonably long, precise and relevant time series of measurements. As pointed out by Hall and Anderson (2002), uncertainty in extreme or unrepeatable events is more important than in situations where there are historical data set, and it requires different approaches towards uncertainty estimation. The lack of sufficient historical data makes the uncertainty results from the model unreliable.

Another problem which is related to extrapolation is that the MLUE method is applicable only to systems whose physical characteristics do not change with time. In other words, the results are not reliable if the physics of the catchment (e.g., land use change) and hydrometrological conditions (due to climate change) differ substantially from what was observed in calibrating the model. If there is evidence of such changes, then the models should be re-calibrated.

The reliability and accuracy of the uncertainty analysis depend on the accuracy of the uncertainty models used, so attention should be given to these aspects. The proposed method does not consider the uncertainty associated with the model U. For example, one could use the cross-validation data set to improve the accuracy of the model U by generalising its predictive capability.

Chapter 5
Application of Machine Learning Method to Predict Parameter Uncertainty

---

This chapter presents the application MLUE (machine learning method to predict parameter uncertainty) method to rainfall-runoff models. We use three machines learning methods, namely artificial neural networks, model trees; and locally weighted regression to predict several uncertainty descriptors of the rainfall-runoff model outputs. The generalised likelihood uncertainty estimation (GLUE) method is first used to assess the parameter uncertainty of the model and then machine learning methods are used to predict the uncertainty estimated by the GLUE method. The Hydrologiska Byråns Vattenbalansavdelning (HBV) conceptual rainfall-runoff model of the Brue catchment in the United Kingdom is used for the application of the MLUE method.

5.1 Description of the Brue catchment [1]

The Brue catchment is located in South West of England, UK. This catchment has been extensively used for research on weather radar, quantitative precipitation forecasting and rainfall-runoff modelling, as it has been facilitated with a dense rain gauges network as well as coverage by three weather radars. Numerous studies (Bell and Moore, 2000; Moore, 2002) have been conducted regarding the catchment, notably during the period of the Hydrological Radar EXperiment (HYREX) which was a UK Natural Environment Research Council (NERC) Special Topic Program. Figure 5.1 shows the locations of the Brue catchment and the gauging stations. The major land use is pasture on clay soil and there are some patches of woodland in the higher eastern part of the catchment.

---

[1] Based on: Shrestha, D. L., Kayastha, N. and Solomatine, D. (2009). A novel approach to parameter uncertainty analysis of hydrological models using neural networks. *Hydrology and Earth System Sciences*, 13, 1235-1248.

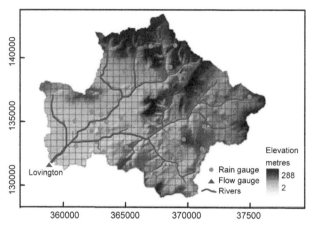

Figure 5.1. The Brue catchment showing dense rain gauges network (the horizontal and vertical axes refer to the easting and northing in British national grid reference co-ordinates).

The catchment has a drainage area of 135 km^2 with the average annual rainfall of 867 mm and the average river flow of 1.92 m^3/s, for the period from 1961 to 1990. Besides weather radar, there is a dense rain gauges network which comprises 49 Cassella 0.2 mm tipping-bucket rain gauges, having recording time resolution of 10 seconds (Bell and Moore, 2000). The network provides at least one rain gauge in each of the 2 km grid squares that lie entirely within the catchment. The discharge is measured at Lovington.

The hourly data of rainfall, discharge, and automatic weather data (temperature, wind, solar radiation etc.) were computed from the 15 minute data. The basin average rainfall data was used in the study. The hourly potential evapotranspiration was computed using the modified Penman method recommended by FAO (Allen et al., 1998). One year hourly data from 1994/06/24 05:00 to 1995/06/24 04:00 was selected for calibration of the HBV model and data from 1995/06/24 05:00 to 1996/05/31 13:00 was used for the verification (validation or testing) of the HBV model. Each of the two data sets represents almost a full year of observations. Statistical properties of the discharge data are shown in Table 5.1.

Table 5.1. Statistical properties of the discharge data.

Statistical properties	Available data	Calibration set	Verification set
Period (yyyy/mm/dd hh:mm)	1994/06/24 05:00 – 1996/05/3113:00	1994/06/24 05:00 – 1996/05/31 13:00	1994/06/24 05:00 – 1996/05/31 13:00
Number of data	16977	8760	8217
Average (m^3/s)	1.91	2.25	1.53
Minimum (m^3/s)	0.15	0.17	0.14
Maximum (m^3/s)	39.58	39.58	29.56
Standard deviation (m^3/s)	3.14	3.68	2.37

5.2 Description of rainfall-runoff model

In this thesis we use simplified version of the HBV (Bergström, 1976) model as rainfall-runoff model. The HBV model is a lumped conceptual hydrological model which includes conceptual numerical descriptions of the hydrological processes at catchment scale. The model was developed at the Swedish Meteorological and Hydrological Institute (Bergström, 1976) and is named after the abbreviation of Hydrologiska Byråns Vattenbalansavdelning (Hydrological Bureau Water Balance Section). The model was originally developed for Scandinavian catchments, but has been successfully applied in more than thirty countries all over the world (Lindström et al., 1997). With growing application of the HBV model, there are also interests to quantify the prediction uncertainty of the HBV models (see, e.g., Seibert, 1997; Uhlenbrook et al., 1999).

The simplified version of the HBV model follows the structure of the HBV-96 model (Lindström et al., 1997) and its schematic diagram is shown in Figure 5.2. The model comprises subroutines for snow accumulation and melt, soil moisture accounting procedure, routines for runoff generation, and a simple routing procedure. The snowmelt routine is based on a degree day relation with an altitude correction for precipitation and temperature:

$$snowmelt = CFMAX(T - TT) \tag{5.1}$$

where TT is the threshold temperature, T is the altitude corrected temperature; and the parameter $CFMAX$ is the melting factor. The threshold temperature is usually close to 0^0 C and is used to define the temperature above which snowmelt occurs. The threshold temperature is also used to decide whether the precipitation falls as rain or snow. If the mean air temperature is less than the threshold temperature, precipitation is assumed to

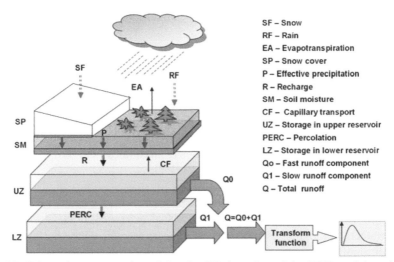

Figure 5.2. Schematic representation of the simplified version of the HBV model used in this thesis with routines for snow, soil, and runoff response (adapted from Shrestha and Solomatine, 2008).

be in snow form. The snow pack is assumed to retain melt water as long as the amount does not exceed a certain fraction (given by the parameter *WHC*) of the snow. When temperature decreases below the threshold temperature, this water refreezes according to the formula:

$$refreezing\ meltwater = CFR \cdot CFMAX(TT - T) \qquad (5.2)$$

where *CFR* is the refreezing factor.

The soil moisture accounting routine computes the proportion of snowmelt or rainfall *P* (mm/h or mm/day) that reaches the soil surface, which is ultimately converted to runoff. This proportion is related to the soil moisture deficit and is calculated using the relation (see also Figure 5.3a):

$$\frac{R}{P} = \left(\frac{SM}{FC}\right)^{BETA} \qquad (5.3)$$

where *R* is the recharge to the upper zone (mm/h or mm/day), *SM* is the soil moisture storage (mm), *FC* is the maximum soil moisture storage (mm), and *BETA* is a parameter accounting for non linearity. If the soil is dry (i.e., small value of *SM*/*FC*), the recharge *R*, which subsequently becomes runoff, is small as the major portion of the effective precipitation *P* is used to increase the soil moisture. Whereas if the soil is wet, the major portion of *P* is available to increase the storage in the upper zone.

Actual evapotranspiration *EA* (mm/h or mm/day) from the soil moisture storage is calculated from the potential evapotranspiration *EP* (mm/h or mm/day) using the following formulae (see Figure 5.3b):

$$EA = EP\left(\frac{SM}{FC \cdot LP}\right) \quad \text{if } SM < FC \cdot LP \qquad (5.4)$$

$$EA = EP \qquad\qquad\quad \text{if } SM \geq FC \cdot LP$$

where *LP* is the fraction of *FC* above which the evapotranspiration reaches its potential level. The actual evapotranspiration that takes place from the soil moisture storage depends on the soil moisture. Evapotranspiration is equal to the potential value if the relative soil moisture (i.e., *SM*/*FC*) is greater than *LP*. If the relative soil moisture is less than this value, the actual evapotranspiration is reduced linearly to zero for a completely dry soil.

Runoff generation routine transforms excess water *R* from the soil moisture zone to the runoff. The routine consists of two conceptual reservoirs arranged vertically one over the other. The upper reservoir is a non-linear reservoir whose outflow simulates the direct runoff component from the upper soil zone, while the lower one is a linear reservoir whose outflow simulates the base flow component of the runoff. Excess water or recharge *R* enters the upper reservoir and its outflow is given by:

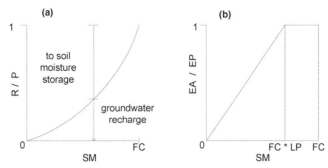

Figure 5.3. HBV model parameters relations. (a) Contributions from precipitation to the soil moisture or ground water storage and (b) ratio of actual and potential evapotranspiration.

$$Q_0 = K \cdot UZ^{(1+ALFA)} \tag{5.5}$$

where K is recession coefficient of the upper reservoir, UZ is the storage in the upper reservoir (mm), $ALFA$ is the parameter accounting for the non-linearity. There is also a capillary flux CF (mm/h or mm/day) from the upper reservoir to the soil moisture zone and is computed by the following formula:

$$CF = CFLUX\left(1 - \frac{SM}{FC}\right) \tag{5.6}$$

where $CFLUX$ is the maximum value of capillary flux. The lower reservoir is filled by a constant percolation rate $PERC$ (mm/h or mm/day), as long as there is water in the upper reservoir. Outflow from the lower reservoir is calculated according to the following equation:

$$Q_1 = K_4 \cdot LZ \tag{5.7}$$

where K_4 is the recession coefficient of the lower reservoir, LZ is the storage in the lower reservoir (mm). The total runoff Q is computed as the sum of the outflows from the upper and the lower reservoirs. The total runoff is then smoothed using a triangular transformation function whose base is defined by a parameter $MAXBAS$ (hours or days).

Input data are observations of precipitation, air temperature and estimates of potential evapotranspiration. The time step is usually one day, but it is possible to use shorter time steps. The evaporation values used are normally monthly averages, although it is possible to use the daily values. Air temperature data are used for the calculations of snow accumulation and melt. It can also be used to adjust potential evapotranspiration when the temperature deviates from normal values, or to calculate the potential evapotranspiration. A detailed description of the HBV-96 model can be found elsewhere (see, e.g., Lindström et al., 1997).

5.3 Experimental setup

The simplified version of the HBV model consists of 13 parameters (4 parameters for snow, 4 for soil, and 5 for the response routine). Since there is little or no snowfall in the catchment, the snow routine was excluded leaving only 9 parameters (see Table 5.2). The model is first calibrated using the global optimisation routine – adaptive cluster covering algorithm, ACCO (Solomatine et al., 1999) to find the best set of parameters, and subsequently manual adjustments of the parameters are made. ACCO is a random search global optimisation algorithm which is implemented in the global optimisation tool, GLOBE (available at http://www.data-machine.com). Briefly the ACCO algorithm works as follows:

1. The initial population of parameters (points) is sampled uniformly in the feasible domain;

2. The population is reduced by choosing the "good" points (promising points with the low function value);

3. The good points are grouped into clusters; and

4. For each cluster, smaller regions are repetitively formed around the currently best point and progressively covered by the newly generated points.

A detailed description of the algorithm can be found in the citation above.

The ranges of parameters values for automatic calibration and uncertainty analysis are set based on the ranges of calibrated values from the other model applications (e.g., Braun and Renner, 1992) and the hydrological knowledge of the catchment. The ranges are extended when the solutions are found near the border of the parameter ranges and re-calibration of the model is done with the extended ranges of the parameters.

Table 5.2. Ranges and calibrated values of the HBV model parameters.

Parameter	Description and unit	Ranges	Calibrated value
FC	Maximum soil moisture content (mm)	100-300	160.335
LP	Ratio for potential evapotranspiration (-)	0.5-0.99	0.527
ALFA	Response box parameter (-)	0-4	1.54
BETA	Exponential parameter in soil routine (-)	0.9-2	1.963
K	Recession coefficient for upper tank (/hour)	0.0005-0.1	0.001
K4	Recession coefficient for lower tank (/hour)	0.0001-0.005	0.004
PERC	Maximum flow from upper to lower tank (mm/hour)	0.01-0.09	0.089
CFLUX	Maximum value of capillary flow (mm/hour)	0.01-0.05	0.0038
MAXBAS	Transfer function parameter (hour)	8-15	12

Note: The uniform ranges of parameters are used for calibration of the HBV model using the ACCO algorithm and for analysis of the parameter uncertainty of the HBV model.

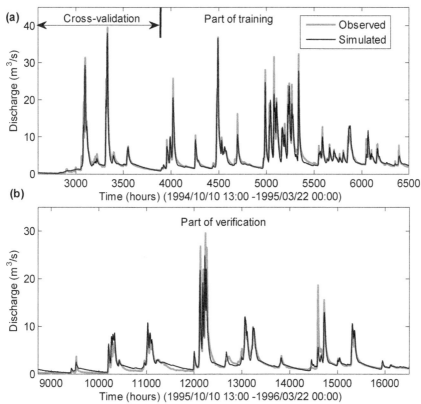

Figure 5.4. Simulated discharge for the Brue catchment in a part of (a) calibration period and (b) verification period.

Sometimes automatic calibration gives the parameter values which do not represent the physical process well in all situations. Therefore, manual fine tuning of the parameters follows the automatic procedure by visual comparison of the observed and simulated hydrographs

The model is calibrated using the Nash-Sutcliffe model efficiency, CoE (Nash and Sutcliffe, 1970) as a performance measure of the HBV model. The CoE value of 0.96 is obtained for the calibration period. The model is validated by simulating the flows for the independent verification data set, and the CoE is 0.83 for this period. Figure 5.4 shows the observed and simulated hydrograph in a part of calibration and in verification period. The scatter plot of the observed and simulated hourly flow is presented on Figure 5.5. For many data points HBV model is quite accurate but its error (uncertainty) is quite high during the peak flows. This can be explained by the fact that the version of the HBV model used in this study is the lumped model and one cannot expect high accuracy from it. Furthermore input data might be erroneous. It is observed that residuals are autocorrelated and heteroscedastic (Shrestha and Solomatine, 2008).

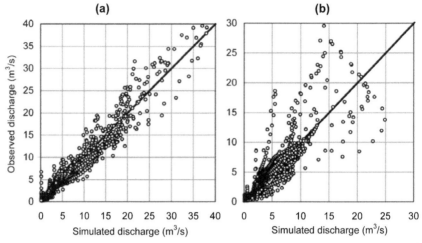

Figure 5.5. Scatter plot of observed versus simulated hourly discharge by HBV model. (a) Calibration period and (b) verification period.

5.4   MC simulations and convergence analysis

The parameters are sampled from the uniform distribution with the ranges given in Table 5.2. The model is run for each random parameter set and the likelihood measure is computed for each model run. We use the sum of the squared errors as the basis to calculate the likelihood measure (see Freer et al., 1996) in the form:

$$L(\theta_i / D) = (1 - \frac{\sigma_e^2}{\sigma_{obs}^2})^N \qquad (5.8)$$

where $L(\theta_i/D)$ is the likelihood measure for the ith model conditioned on the observations D, σ_e^2 is the associated error variance for the ith model, σ_{obs}^2 is the observed variance for the period under consideration, N is a user defined parameter. We set N to 1; in this case equation (5.8) is equivalent to CoE.

We investigate the number of behavioural samples retained out of 74,467 MC samples for different values of the rejection threshold. Figure 5.6 shows the percentage of behavioural samples corresponding to different values of the rejection threshold as measured by CoE. It is observed that only about 1/3 of simulations (25,000 samples) are accepted for the threshold value of 0, whereas less than 1/10 of simulations are retained for a higher threshold value of 0.7. Based on total samples of 74467, we also compute the ratio of total number of samples and number of behavioural samples for different values of the rejection threshold. The results show that the ratio increases exponentially with the threshold value. These results suggest that the MC sampling within the GLUE method is not an efficient sampling method as the larger number of samplings are rejected.

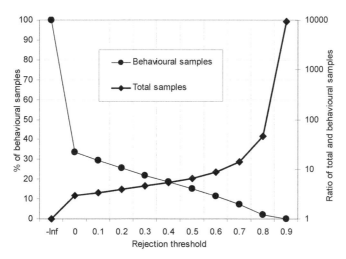

Figure 5.6. Percentage of behavioural samples (left y axis) and ratio of total number of samples and behavioural samples (right y axis) with different value of the rejection threshold as measured by the coefficient of model efficiency, CoE.

The convergence of the MC simulations is also assessed to know the number of samples required to obtain reliable results. Since we have used CoE as an objective function to calibrate the model and as a likelihood measure in the GLUE method, we use the same metric also to determine the convergence of the MC simulations. The mean and standard deviation of CoE are used to analyse the convergence of the MC simulations:

$$ME_k = \frac{1}{k}\sum_{i=1}^{k} CoE_i$$

(5.9)

$$SDE_k = \sqrt{\frac{1}{k}\sum_{i=1}^{k}(CoE_i - ME_k)^2}$$

where CoE_i is the coefficient of model efficiency of the ith MC run, ME_k and SDE_k are the mean and standard deviation of the model efficiency up to the kth run, respectively.

The Figure 5.7 depicts the two statistics – the mean and standard deviation of the CoE of model runs corresponding to different numbers of samples. It is observed that both statistics become stable around 10,000 simulations; however, in this case study we used 25,000 behavioural samples. Note that we check the convergence after getting the 25,000 behavioural samples from the 74,467 MC simulations. But it is better to monitor these statistics during the MC simulations and stop the simulation when the convergence criterion is met. Other statistics such as the coefficients of variation for the mean and variance of the model efficiency can also be used to test the convergence (see, e.g., Ballio and Guadagnini, 2004).

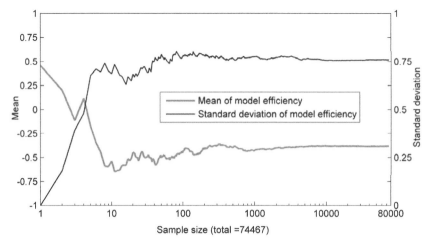

Figure 5.7. The convergence of mean and standard deviation of Nash-Sutcliffe model efficiency. Note that x-axis is log scale to see initial variation.

We also computed several other performance metrics (e.g., RMSE, absolute error, percentage bias and others) for the convergence test. It is observed that the results are consistent with the CoE metric. Furthermore we also analysed the convergence of the runoff predictions at some arbitrary points (e.g., in peak, medium and base flow), and the results are quite consistent with the previous global performance metrics.

5.5 Posterior distributions and sensitivity of parameters

Figure 5.8 shows a scatter plot (also called dotty plot) of the 74,465 MC samples associated with the likelihood measure of equation (5.8). This figure shows that there are many combinations of parameters values for a chosen model structure that may be equally good in reproducing the observed discharge in terms of some quantitative objective function of goodness of fit. It also appears that the good simulations are distributed across a wide range of values for any particular parameter, suggesting that it is the combined set of parameters that is important (see, Freer et al., 1996). If the surface of the scatter plot (representing the best model parameterisation for a given value of the parameter) has a clearly defined minimum or maximum then the parameter can be considered to be well identified. In our case study only two parameters *ALFA* and *MAXBAS* emerge as identifiable parameters.

Figure 5.9 shows the marginal posterior probability distributions of the model parameters resulting from the MC simulation for a period of 2 years of the hourly stream flow data. These distributions are conditioned on the given model structure and the likelihood function. It is noteworthy to mention that these distributions are not derived from the Bayesian statistics but they are approximated from MC samples. So these distributions are not truly posterior distribution and will be referred to as empirical posterior distributions. These distributions are derived from the likelihood measure of equation (5.8).

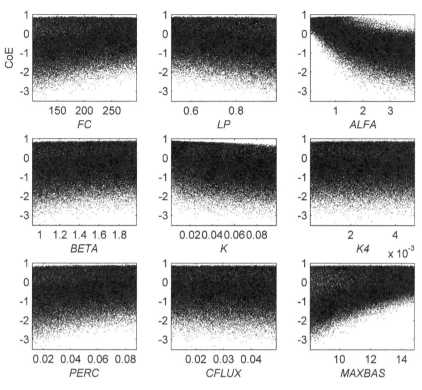

Figure 5.8. Dotty plots resulting from MC simulations. Each dot represents one model run with the parameter vector chosen randomly by uniform sampling across the range for each parameter.

The marginal empirical posterior distribution for each parameter $\theta^{(j)}$ from the joint probability distribution of the parameter vector $\theta_i^{(j)}$, $i=1, .., s$ and $j=1, ..., p$ (where s is the number of samples and p is the number of parameters) and the corresponding likelihood value L_i is computed by the following procedure:

1. For each parameter $\theta^{(1)}$, divide the range of its sampled values $\theta_i^{(1)}$ obtained from the MC samples into K equal bins;

2. For each bin k ($1 \leq k \leq K$),

   • Compute the sampling density $nsamp_k$ by counting the number of samples of the parameter $\theta^{(1)}$ belonging to the bin k;

   • Sum the likelihood values L_i of all $nsamp_k$ samples which belong to the bin k. This sum is denoted by L_k;

   • Compute the probability $p_k = L_k / nsamp_k$;

3. Repeat step 2 for all bins;

4. Normalize the probability p_k to have the sum of the probability one: $p_k = \dfrac{p_k}{\displaystyle\sum_{k=1}^{K} p_k}$; and

5. Repeat steps 1-4 for all parameters.

The figure shows that some parameters such as *K4* and *CFLUX* have relatively flat distributions while some parameters like *ALFA*, *K*, and *MAXBAS* have skewed distributions. An additional analysis of the posterior distributions with different values of rejection threshold has been carried out and the results reveal that the posterior distributions are sensitive to the rejection threshold value.

Figure 5.10 shows the cumulative distributions of the best performing (with respect to the likelihood value) 10% parameter sets. The cumulative distributions for each parameter are computed following the same procedure given for the marginal posterior distribution. The only difference is that instead of using all the MC samples, the best 10% parameter sets are selected. The gradient of the cumulative distribution measures the magnitude of the sensitivity of the parameters. The higher gradient of the distribution such as in parameters *ALFA* and *K* means a high sensitivity of the parameter.

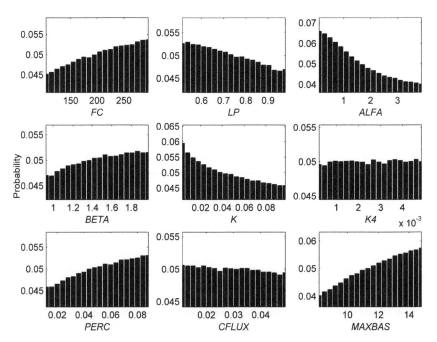

Figure 5.9. Empirical posterior probability distribution of 74467 numbers of sample parameters condition on likelihood based on the coefficient of model efficiency. Note that the prior distribution for these parameters is uniform.

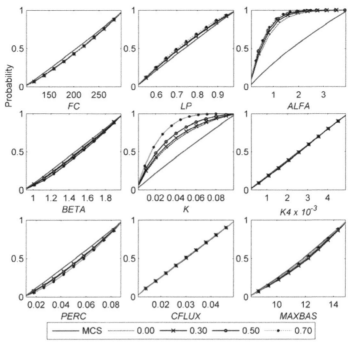

Figure 5.10. Sensitivity plot of the best 10% parameter sets of MC simulations for different values of CoE rejection threshold (0, 0.3, 0.5, 0.7).

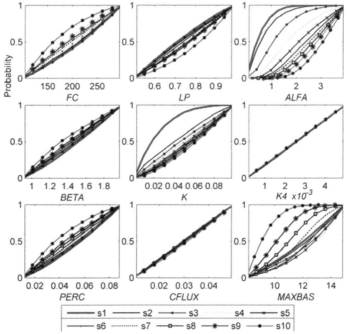

Figure 5.11. Sensitivity plot of parameters expressed as cumulative distributions of values in 10 equal sets (by number). Set s1 represents the highest likelihood values and set s10 the lowest.

Further analysis is carried out by computing the cumulative distributions of the behavioural samples for different values of the rejection threshold. It is observed that there are three groups of the parameters: (i) insensitive parameters such as *K4* and *CFLUX*; (ii) those with a positive gradient such as *FC*, *LP*, *K*, and *MAXBAS*; and (iii) those with a negative gradient such as *ALFA*, *BETA*, *K4*, and *PERC*. Parameters from the first group are insensitive to the threshold values and can be considered as poorly identifiable parameters. In the second group of parameters, the gradient increases with the increase of the threshold value. In the third group of parameters, the gradient decreases with the increase of the threshold value. Such plots must be interpreted with care, and the visual impression will depend on the original range of the parameter considered, and the likelihood function used. It should be noted that parameters which show the uniform marginal distribution may still have significance in the context of a set of values of the other parameters due to the parameter correlation and interaction.

As shown in Figure 5.8, the interaction between parameter values results in broad regions of acceptable simulations and it is the combination of parameter values that produces the acceptable or non-acceptable simulations within the chosen model structure. Such a plot reveals little about the sensitivity of the model predictions to the individual parameters, except where some strong change in the likelihood measure is observed in some range of a particular parameter. A more detailed analysis based on the extension of regional sensitivity analysis (Spear and Hornberger, 1980; Freer et al., 1996) was performed. Figure 5.11 shows the cumulative distributions for 10 equal sets (by number) of behavioural simulations corresponding to the rejection threshold value of 0. Each parameter population is ranked from best to worst in terms of the chosen likelihood function, and the ranked population is then divided into ten bins of equal size (2500 numbers of parameter sets). The cumulative likelihood distribution of each group is then plotted. The parameter sensitivity can be evaluated by assessing the spread of the cumulative distribution function for each group. Parameters with a strong variation in the distribution function are recognized as sensitive parameters. *ALFA*, *K* and *MAXBAS* are such sensitive parameters, while *K4* and *CLUX* show no sensitivity at all. The rest of the parameters show a moderate sensitivity.

5.6 Machine learning techniques in emulating results of Monte Carlo simulations

MC simulation provides the basis for assessing the model parameter uncertainty. Once the model uncertainty is estimated, a machine learning model can be trained to learn the functional relationship between the uncertainty results and the input data. In the considered case study the GLUE method has been used for the parameter uncertainty estimation of the HBV model. The threshold value of 0 (measured by CoE) is selected to classify the simulation as either behavioural or non-behavioural. 90% uncertainty bounds are calculated using the 5% and 95% quantiles of the MC simulation realizations weighted with corresponding likelihood values (equation (4.5)).

The corresponding 90% lower and upper PI (see equation (4.7)) are calculated using the model output simulated by the optimal parameter sets shown in Table 5.2. Hence the computed upper and lower PI is conditioned by the optimal model output. The next important step is to select the most relevant input variables to build the machine learning models from the domain of the input data.

5.6.1   Selection of input variables

In order to select the most important influencing variables for the machine learning model, several approaches can be used (see, e.g., Guyon and Elisseeff, 2003; Bowden et al., 2005). We follow the approaches discussed in section 4.6. The input variables for the machine learning model U can be constructed from the input such as rainfall, evapotranspiration, and the output such as observed discharge of the primary model (the HBV model). Experimental results show that the evapotranspiration alone does not have a significant influence on the PIs of the model output. Thus it is decided not to include the evapotranspiration as a separate variable, but rather to use effective rainfall. The effective rainfall denoted by RE_t is computed from:

$$RE_t = \max((R_t - EP_t), 0)$$

(5.10)

where R_t is the rainfall and EP_t is the evapotranspiration.

Figure 5.12 shows the correlation coefficient and the AMI of RE_t and its lagged variables with the lower and upper PIs. It is observed that the correlation coefficient is smallest at a zero hour lag time and increases as the lag increases up to 7 hours (Figure 5.12b) for the upper PI. The correlation plot (see Figure 5.12a) for the lower PI is different than the upper PI. The optimal lag time (the time at which the correlation coefficient and/or AMI is maximum) is 9 hours. Such findings are also supported by the AMI analysis. At the optimal lag time, the variable RE_t provides the maximum amount of information about the PIs.

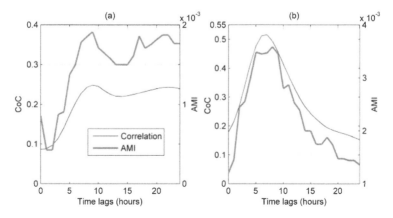

Figure 5.12. Linear correlation and average mutual information (AMI) between effective rainfall (a) lower prediction interval and (b) upper prediction interval for different time lags.

Table 5.3. Input data structures of machine learning models to reproduce Monte Carlo uncertainty results of the HBV model.

Models	Lower prediction interval	Upper prediction interval
V11	RE_{t-9a}, Q_{t-1}, ΔQ_{t-1}	RE_{t-9a}, Q_{t-1}, ΔQ_{t-1}
V22	RE_{t-7a}, Q_{t-1}, ΔQ_{t-1}	RE_{t-7a}, Q_{t-1}, ΔQ_{t-1}
V33	RE_{t-7}, RE_{t-8}, RE_{t-9}, Q_{t-1}, ΔQ_{t-1}	RE_{t-7}, RE_{t-8}, RE_{t-9}, Q_{t-1}, ΔQ_{t-1}
V44	RE_{t-9a}, Q_{t-1}, Q_{t-2}	RE_{t-9a}, Q_{t-1}, Q_{t-2}
V12	RE_{t-9a}, Q_{t-1}, ΔQ_{t-1}	RE_{t-7a}, Q_{t-1}, ΔQ_{t-1}

Additionally, the correlation and AMI between the PIs and the observed discharge are analysed. The results show that the immediate and the recent discharges (with the lag of 0, 1, and 2 hours) have a very high correlation with the PIs. So it is also decided to use the past values of the observed discharge as additional input to the model U.

Based on the above analysis, several structures for the input data for the machine learning models are considered. The principle of parsimony is followed to avoid the use of a large number of inputs, so the aggregates, such as moving averages or derivatives of the inputs that would have a hydrological meaning were considered. Typically, the rainfall depth at a shorter (e.g., an hourly) time step partially exhibits a random behaviour, and is not very representative of the rainfall phenomenon during the short period. Considering the maximum dependence to 7 hours and 9 hours lag time, we use two mean effective rainfall values:

RE_{t-7a} - the mean of RE_{t-3}, RE_{t-4}, RE_{t-5}, RE_{t-6}, RE_{t-7}; and
RE_{t-9a} - the mean of RE_{t-5}, RE_{t-6}, RE_{t-7}, RE_{t-8}, RE_{t-9}.

Furthermore, the derivative of the flow indicates whether the flow situation is either normal or base flow (zero or small derivative), or can be characterized as the rising limb of the flood event (high positive derivative), or the recession limb (high negative derivative). Therefore, in addition to the flow variable Q_{t-1}, the rate of flow change ΔQ_{t-1} at time $t-1$ is also considered, where $\Delta Q_{t-1} = Q_{t-1} - Q_{t-2}$.

Table 5.3 presents five possible combinations of input data structure considered for the machine learning models. Note that these two models are trained independently for producing the lower and upper PI; however, it is possible to use a single model in some machine learning techniques (e.g., ANN) to produce two outputs simultaneously (in this case the model uses the same input data).

The structure of the two machine learning models to estimate the lower and upper PI, for instance in V11 input configuration, takes the following form:

$$PI_t^L = U_L(RE_{t-9a}, Q_{t-1}, \Delta Q_{t-1}) \tag{5.11}$$

$$PI_t^U = U_U(RE_{t-9a}, Q_{t-1}, \Delta Q_{t-1})$$

where, PI_t^L and PI_t^U are the lower and upper PI for tth time step. Note that we did not include the variable Q_t to the input of the machine learning model above because during the model application this variable is not available (indeed, the prediction of this variable is done by the HBV model and the machine learning model assesses the uncertainty of the prediction). Furthermore, we would like to stress that, in this study the uncertainty of the model output is assessed when the model is used in simulation mode. However, this method can also be used in forecasting mode, provided that the primary model is also run in forecasting mode.

5.6.2 Modelling prediction intervals

In this study, we use three machine learning models, namely artificial neural networks (ANNs), model tree (MT), and locally weighted regression (LWR). The same data sets used for the calibration and verification of the HBV model are used for the training and verification of the model U, respectively. However, for proper training of the machine learning models the calibration data set is segmented into two subsets: 15% of data sets for cross-validation (CV) and 85% for training (see Figure 5.4). The CV data set is used to identify the best structure of the machine learning models.

Artificial neural networks

A multilayer perceptron network with one hidden layer is used; optimization is performed by the Levenberg-Marquardt algorithm. The hyperbolic tangent function is used for the hidden layer with the linear transfer function at the output layer. The maximum number of epochs is fixed at 1000. The trial-and-error method is adopted to detect the optimal number of neurons in the hidden layer, testing a number of neurons from 1 to 10. It is observed that seven and eight neurons for the lower and upper PI, respectively give the lowest error on the CV data set.

The performance of the ANN based uncertainty models U for different input structures is shown in Table 5.4. All ANN models are characterized by similar values of the coefficient of correlation (CoC) in the CV data set producing a lower PI. In the verification data set, V11 or V12 (note that configurations V11 and V12 have the same ANN model for the lower PI), and V44 have the highest CoC value. In producing the upper PI, V22 or V12 (configurations V22 and V12 have the same ANN model for the upper PI), and V44 give the highest CoC value in the CV data set.

Table 5.4. Performances of ANN models measured by the coefficient of correlation (CoC), the prediction interval coverage probability (PICP) and the mean prediction interval (MPI).

Models	CoC for lower PI			CoC for upper PI			PICP	MPI
	Training	CV	Verification	Training	CV	Verification	(%)	(m^3/s)
V11	0.91	0.95	0.86	0.80	0.86	0.80	77.00	2.09
V22	0.92	0.95	0.83	0.83	0.88	0.79	81.78	2.24
V33	0.91	0.95	0.79	0.75	0.84	0.76	73.70	1.90
V44	0.90	0.96	0.86	0.79	0.88	0.80	73.99	1.93
V12	0.91	0.95	0.86	0.83	0.88	0.79	76.55	2.09

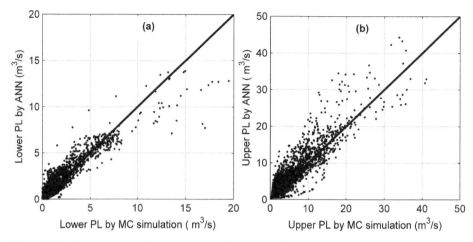

Figure 5.13. Scatter plots showing the performance of the ANN model to reproduce 90% prediction limits (PL) of MC simulations in the verification data set. (a) 90% lower PL and (b) 90% upper PL. X-axes show the PLs obtained by MC simulations and Y-axes show the PL estimated by ANN.

Figure 5.13 presents the scatter plots of one of the ANN models to produce the lower and upper prediction limits in the verification period. It is recalled that prediction limits are computed by adding the model output to the PIs estimated by ANN models (see equation (4.12)). It appears that the CoC values obtained when predicting the lower prediction limits are higher than those for the upper prediction limits. This can be explained by the fact that the upper prediction limits correspond to the higher values of flow (where the HBV model is less accurate) and have higher variability, which makes the prediction a difficult task.

We also compared the percentage of the observation data falling within the PIs (i.e., PICP) produced by the MC simulations and the ANN models. The value of PICP obtained by the MC simulations is 77.24%. Note that we specified the confidence level of 90% to produce these PIs and theoretically one would expect a PICP value of 90%. However it has been reported that the PICP obtained by the MC simulations is normally much lower than the specified confidence level used to produce these PIs. This low value of PICP is consistent with the results reported in the literature (see, e.g., Montanari, 2005; Xiong and O'Connor, 2008). The low efficiency of the PIs obtained by the MC simulations in enveloping the real-world discharge observations might be mainly due to the following three reasons among others:

- The uncertainty in the model structure, the input (such as rainfall, temperature data) and output discharge data are not considered in the MC simulations.

- We assumed a uniform distribution and ignored the parameter correlation. These assumptions could be wrong.

- We employed the GLUE method and the width of the uncertainty bound obtained by this method varies with the rejection threshold and the likelihood measure to a great extent.

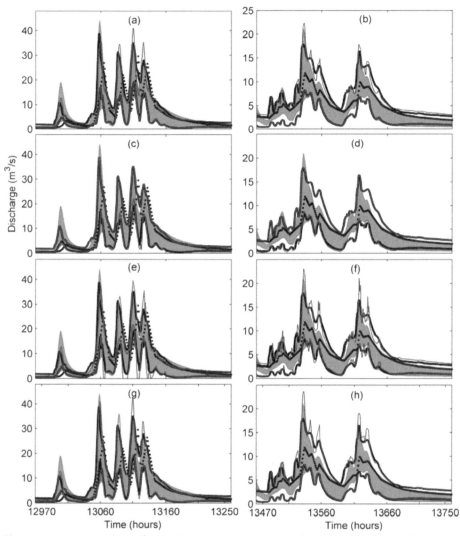

Figure 5.14. A comparison of 90% PIs estimated by the MC simulations (darker shaded region) and V12 (thick lines) with (a and b) V11, (c and d) V22, (e and f) V33, (g and h) V44 in parts of the verification period. The black dots indicate observed discharges and the blue lines denote 90% PIs obtained by V11, V22, V33, V44.

Among the ANN models, V22 gives the best result corresponding to a PICP value (i.e. 81.78% against the required 90%). However, V11 and V12 give PICP values very close to the MC simulation results. The average width of the PIs (i.e., MPI) by V22 is wider; this is one of the reasons to include a little bit more observed data inside the PIs. V33 gives the lowest values of both PICP and MPI. The MPI values of V11 and V12 are very close (almost equal) to the MC simulation results.

Figure 5.14 shows the comparison of the 90% PIs estimated by the MC simulations with 5 different ANN configurations of input data in the verification period. We also compare the PIs of V12 with those of V11 (Figure 5.14a and b), V22 (Figure 5.14c and d), V33 (Figure 5.14e and f) and V44 (Figure 5.14g and h). It is observed that the upper PIs on peak flows are overestimated by V11 compared to V12. Note that the lower PIs of V11 and V12 coincide because the input structures of the V11 and V12 for the lower PIs are the same. One can see from Figure 5.14c and d that V22 and V12 give similar results with a slight underestimation of the lower PI. In this comparison, the upper PI of V22 and V12 coincide because of the same input structures for the upper PI.

One can see a noticeable difference between V33 and V12 for both the upper and the lower PI (see Figure 5.14e and f). Most of the upper PIs on the peaks are overestimated by V33 and the lower PIs on most of peaks are considerably underestimated. This result is not surprising if one analyses the structure of the input data. The obvious reason is that the input data consists of instantaneous values of RE_{t-7}, RE_{t-8}, RE_{t-9}, while in the other models we use a moving average value of effective rainfall, and thus the smooth function of the ANN is produced. Figure 5.14g and h show that inclusion of the input ΔQ_{t-1} does not improve the accuracy as expected. However overestimation of the upper PIs on peaks is due to the combined effect of considering RE_{t-9} (while upper PI has maximum correlation at 7 hours) and of not using ΔQ_{t-1}.

From the above analysis it can be concluded that model V12 is relatively better: it produces 90% PIs close to the 90% PIs obtained by the MC simulations. It can be said that in general most of the ANN models (except V33) reproduce the MC simulations uncertainty bounds reasonably well except for some peaks, in spite of the low correlation of the input variables with the PIs. Although some errors can be noticed, the predicted uncertainty bounds follow the general trend of the MC uncertainty bounds. Noticeably, the model fails to capture the observed flow during one of the peak events (Figure 5.14a, c, e, and g). Note however, that the results of the ANN models and the MC simulations are visually closer to each other than both of them to the observed data. It is also observed from Figure 5.14 that the model prediction uncertainty caused by parameter uncertainty is rather large. There could be several reasons for this including:

- The GLUE method does not strictly follow the Bayesian inference process (Mantovan and Todini, 2006), which leads to an overestimation of the model prediction uncertainty.

- The uncertainty bound very much depends on the rejection threshold which is used to distinguish between the behavioural and non-behavioural models. In this study we used a quite low value of the rejection threshold (CoE value of 0) which produces relatively wider uncertainty bounds.

- We considered only parameter uncertainty assuming that the model structure and the input data are correct. As mentioned at the beginning of this section, the scatter plot reveals that this assumption is not really correct.

Model trees

Several experiments are carried out with a different number of pruning factors that control the complexity of the generated models. We report the results of the MT which has a moderate level of complexity. Note that the cross-validation data set has not been used in the MT; rather the MT uses all the calibration data set to build the models. Table 5.5 shows the performance of the MT with two different input configurations (see Table 5.3).

It can be observed that V12 gives comparatively better results with respect to the CoC for the upper PI. It is reiterated that the models V11 and V12 to produce the lower PI are the same (because of the same input structures), thus the CoC values are the same. We also compute the CoC value between the lower (or upper) prediction limits of the MT and those of the MC results. It is noteworthy to repeat that the upper and lower prediction limits are computed by adding the optimal model output to the corresponding lower and upper PIs (see equation (4.12)). The CoC values for the lower and upper limits are 0.90 and 0.92, respectively, in the verification period for both of the MT models.

Detailed analysis reveals that 71.11% of the observed data are enclosed within the estimated 90% PIs in the verification period when using the model V11. In the case of the model V12 a slightly lower share (68.72%) of the observed data is inside the 90% PIs. As far as the width of the uncertainty bound is concerned, the MPI values obtained by both models are comparable. If compared to the MC simulation results, the MT gives lower value of the PICP with a slightly lower value of MPI. Note that the PICP and MPI values of the MC simulation results are 77.24% and 2.09 m^3/s, respectively.

Figure 5.15 shows the comparison of the 90% PIs estimated by the MC simulations with 2 different input configurations for the MT in the verification period. It is observed that the upper PIs on some peak flows are overestimated by both models of the MT (i.e. V11 and V12). One can see in the second hydrographs (Figure 5.15b and d) that the lower PIs are underestimated on the peaks. In general, the predicted uncertainty bounds by the MT follow the general trend of the MC uncertainty bounds although some errors can be noticed. Noticeably, the model fails to capture the observed flow during one of the peak events.

Table 5.5. Performances of the MT and LWR measured by the coefficient of correlation (CoC), the prediction interval coverage probability (PICP) and the mean prediction interval (MPI).

ML techniques	Models	CoC for lower PI		CoC for upper PI		PICP (%)	MPI (m^3/s)
		Training	Verification	Training	Verification		
MT	V11	0.91	0.84	0.76	0.79	71.11	1.98
	V12	0.91	0.84	0.81	0.79	68.72	1.95
LWR	V11	0.906	0.822	0.779	0.798	75.16	1.96
	V12	0.906	0.822	0.766	0.744	75.43	1.93

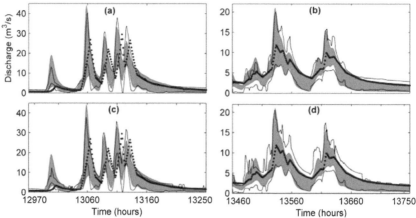

Figure 5.15. Hydrograph of 90% prediction bounds estimated by the MC simulations and the model tree (MT) in verification period with (a and b) V11, (c and d) V12. The black dot indicates observed discharges and the dark grey shaded area denotes the prediction uncertainty that results from the MC simulations. The black line denotes the prediction uncertainty estimated by the MT.

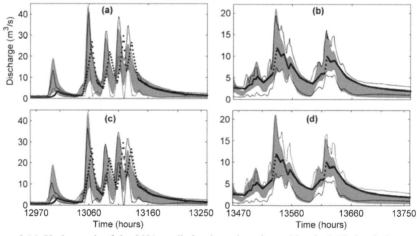

Figure 5.16. Hydrograph of the 90% prediction bounds estimated by the MC simulations and the locally weighted regression (LWR) in the verification period with (a and b) V11, (c and d) V12. The black dot indicates observed discharges and the dark grey shaded area denotes the prediction uncertainty that results from the MC simulations. The black line denotes the prediction uncertainty estimated by the LWR.

Locally weighted regression

We conducted the experiments using LWR. Two important parameters of the LWR are the number of neighbours and the weight function. Several experiments are done with a different combination of these values and we report the best results obtained with 5 neighbours and the linear weight function. The comparison of performance of the LWR with the two different input configurations is shown in Table 5.5.

It can be seen that the model V11 is slightly better than the model V12 with respect to the CoC values in the verification period. The CoC values for the lower and upper limits are 0.892 and 0.923, respectively for the V11 model. For the V12 model they are 0.892 and 0.904, respectively. However, one can see that both models give similar values of the PICP and MPI. Figure 5.16Figure 5.15 shows the comparison of the 90% PIs estimated by the MC simulations with 2 different input configurations in the verification period. The results are comparable with those obtained by the ANN and the MT.

Comparison among ANN, MT and LWR

The predictive capability of different machine learning models in estimating the lower and upper PIs are compared for the verification period and show in Figure 5.17Figure 5.16. All machine learning models used in this case study produce similar performances. In particular, the ANNs are marginally better than the MT and the LWR. Furthermore, by visual inspection of the hydrographs, one can see that the ANN produces smooth uncertainty bounds compared to the other two. We also report the CoC values of the prediction limits, and it appears that the accuracy is increased. It is noteworthy to reiterate that the prediction limits are computed by adding the corresponding optimal model output to the predicted PIs.

Figure 5.18 presents a summary of the comparison statistics (PICP and MPI) of the uncertainty estimation. The ANN model is very close to the MC simulation results with respect to both statistics. The MT and LWR are better than the ANN with respect to MPI (note that a lower MPI is the indication of a better performance), however the prediction limits estimated by them enclose a relatively lower percentage of the observed values as compared to those of the ANN.

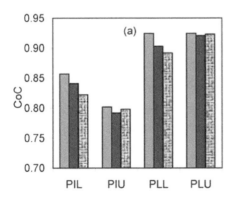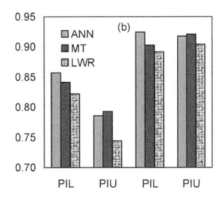

Figure 5.17. Performance of machine learning models measured by coefficient of correlation (CoC) to reproduce the prediction intervals and limits. (a) Input data for upper PI is RE_{t-9a}, Q_{t-1}, ΔQ_{t-1} (i.e. V11 configuration) and (b) input data for upper PI is RE_{t-7a}, Q_{t-1}, ΔQ_{t-1} (i.e. V12 configuration). PIL and PIU denote lower and upper PI, respectively whereas PLL and PLU denote lower and upper prediction limits, respectively.

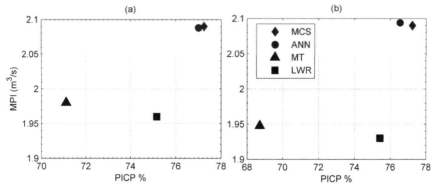

Figure 5.18. A comparison of statistics of uncertainty (PICP and MPI) estimated by the MC simulation, MT, LWR and ANN in the verification period. (a) Input data for the upper PI is RE_{t-9a}, Q_{t-1}, ΔQ_{t-1} (i.e. V11 configuration) and (b) input data for the upper PI is RE_{t-7a}, Q_{t-1}, ΔQ_{t-1} (i.e. V12 configuration).

Table 5.6. Performance criteria of machine learning models indicated by linguistic variables

Models	Accuracy		Efficiency	Transparency	Rank
	CoC	PICP and MPI			
ANN	High	High	Medium	Low	1
MT	Medium	Low	High	Medium	2
LWR	Low	Medium	Low	High	3

So far we compare the performance of three machine learning models through only the accuracy of the prediction; however, there are other factors to be considered as well. These include computational efficiency, simplicity or ease of use, flexibility, transparency etc. They are shown in Table 5.6. We use linguistic variables to describe these factors. It is observed that none of the models is superior with respect to all factors; however one can favour the ANN if the ranking is done by giving equal weight to all factors.

5.6.3 Modelling median and standard deviation

The MLUE method is also used to estimate the median and standard deviation of the MC simulation results. Based on the correlation and AMI analysis, several input data structures are considered. Among them the input data RE_{t-8a}, Q_{t-1}, ΔQ_{t-1} gives better results for predicting the median, while RE_{t-7a}, Q_{t-1}, ΔQ_{t-1} gives better results for predicting the standard deviation. Figure 5.19 shows a comparison of the ANN, MT and LWR to predict the mean and standard deviation of the MC simulation results. One can see that the median is overestimated by all three models except for one of the peaks (Figure 5.19a). The ANN is marginally better than the other two models. As far as the standard deviation is concerned, some peaks are underestimated and some are overestimated. The performance of the machine learning models to predict the standard deviation is a little bit low as compared to the median. Table 5.7 shows the CoC values between the median and standard deviation of the MC simulations and predicted by the three machine learning models.

Table 5.7. Performances of artificial neural networks (ANN), model trees (MT), and locally
weighted regression (LWR) to predict the median and standard deviation of MC simulation.

Models	Coefficient of correlation (CoC)	
	Median	Standard deviation
ANN	0.8953	0.848
MT	0.8903	0.8729
LWR	0.8776	0.8352

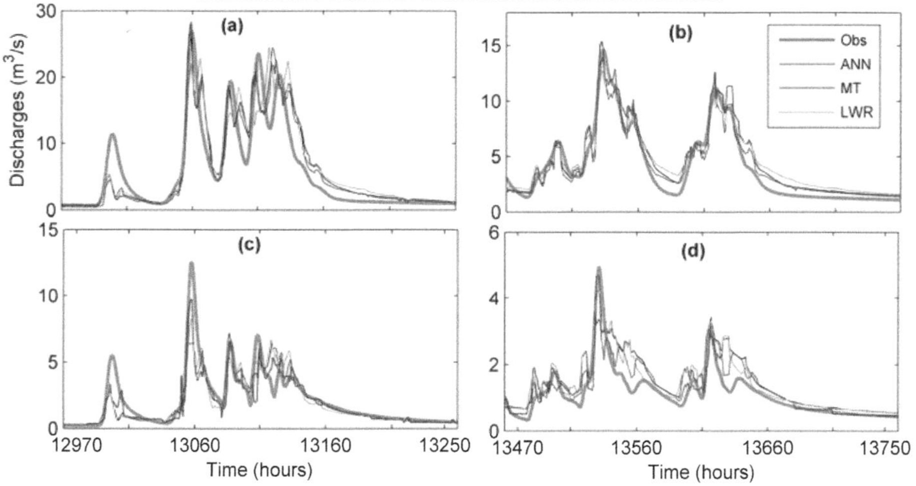

Figure 5.19. A comparison of artificial neural networks (ANN), model trees (MT) and locally
weighted regression (LWR) to predict (a and b) median (c and d) standard deviation of the
MC simulations. Obs is the computed median (a and b) or standard deviation (c and d) of the
MC simulations.

5.6.4 Modelling probability distribution function

In section 5.6.2, we use machine learning methods to estimate the 90% PIs by building
two separate models for the two quantiles (i.e. 5% and 95% quantiles). However it is
possible to extend the methodology to predict several quantiles of the model outputs and
in this way to estimate the distribution functions (pdf or cdf) of the model output
generated by the MC simulations.

The procedure to estimate the pdf of the model output (generated by the MC
simulations) is given below:

1. Derive the cdf of the realizations of the MC simulations in the calibration data;

2. Select several quantiles of the cdf in such a way that these quantile can
   approximate the cdf;

3. Compute the corresponding transferred prediction quantiles using equation
   (4.6);

4. Construct and train separate machine learning models for each transferred
   prediction quantile;

5. Use these models to predict the quantiles for the new input data vector;

6. Construct a cdf from these discrete quantiles by interpolation. This cdf will be an approximation to the cdf of the MC simulations.

In this case study, we select 19 quantiles from 5% to 95% with uniform interval of 5%. Then an individual machine learning model is constructed for each quantile using the same structure of the input data and model that was used in the previous experiments. In principle, we could use a different structure for the input data and the model; this will require additional experiments to achieve the optimal structure. For the sake of simplicity we use the input configuration V11 for all quantiles.

The results are reported in Figure 5.20. The Figure 5.20a shows a comparison of the cdf for the peak event of 1995/12/20 estimated by the three machine learning methods. One can see that the cdf estimated by the ANN, MT and LWR are comparable and are very close to the cdf of the MC simulations. We also compare the cdf for another peak event of 1996/01/09 (see Figure 5.20b). It is observed that the cdf estimated by the ANN, MT and LWR deviate a little bit more near the middle of it. From the visual inspection we have impression that the cdf are reasonably approximated by the machine learning methods. However, it may require a rigorous statistical test to conclude that if the cdf estimated by the machine learning methods are not significantly different from those of the MC simulations. In this case study, since we have limited data (only 19 points), the results of the significant test (e.g., Kolmogorov-Simrnov) may not be reliable.

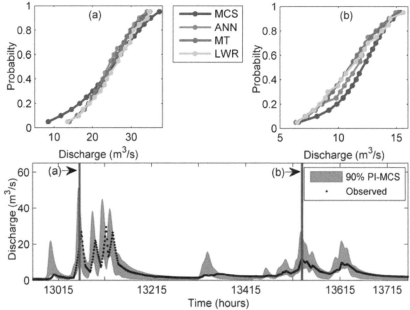

Figure 5.20. A comparison of cumulative distribution function (cdf) estimated with different machine learning method. (a) Peak event of 1995/12/20 and (b) peak event of 1996/01/09.

By visual inspection we can say that the cdf estimated by machine learning methods are reasonably close to those generated by the MCS. It is worth noting that we have not optimised the model and the input data structure of the machine learning methods in the presented results, hence there is hope to improve the results.

One issue of predicting the cdf by the above method is that if the machine learning models are trained independently for several quantiles of the cdf, it cannot be guaranteed that the estimated cdf is a monotonously increasing function. However, if the machine learning models allow the prediction of multiple outputs (e.g., in ANN) then one single model can be built to predict several quantiles simultaneously; and some form of penalty function can be introduced during the training if the cdf is not monotonously increasing. We anticipate future research in this direction.

5.7   Conclusions

This chapter presents the application of MLUE method to predict parameter uncertainty in rainfall-runoff modelling. The Brue catchment (135 km^2) located in South West of England, UK, which has been extensively used for research on weather radar, quantitative precipitation forecasting and rainfall-runoff modelling, is selected for the study. Simplified version of the HBV conceptual model is chosen for the application of the MLUE method. We analysed the parameter sensitivity of HBV model and derived the posterior distribution of its parameters. We identified three groups of the parameters based on the sensitivity analysis.

The GLUE method has been used to analyse the parameter uncertainty of the model. We also investigate the convergence analysis of MC simulation in the GLUE method. Several machine learning models have been applied to predict the uncertainty results generated by the GLUE method. We have shown how domain knowledge and analytical techniques are used to select the input data for the machine learning models used in the MLUE method. Three machine learning models, namely artificial neural networks, model trees, and locally weighted regression, are used to predict the uncertainty of the model predictions. We predict several uncertainty measures such as standard deviation, quantiles; and the pdf of MC realizations. The performance of the MLUE method is measured by its predictive capability (e.g., coefficient of correlation, and RMSE) and the statistics of the uncertainty (e.g., the prediction intervals coverage probability and the mean prediction intervals). It is demonstrated that machine learning methods can predict the uncertainty results with reasonable accuracy. The great advantage of the MLUE method is that once the machine learning models are developed which is done offline, it can predict the uncertainty of the model output in a fraction of second which otherwise would take several hours or days of computation time by the MC based uncertainty analysis methods.

Chapter 6
Machine Learning in Prediction of Residual Uncertainty: UNEEC Method

This chapter presents a novel methodology to analyse, model, and predict the residual uncertainty which, for the purpose of this thesis, is considered to be represented by the model residual errors. The methodology aims to predict the uncertainty of an optimal (calibrated) model output by building machine learning model of the probability distributions of its historical residuals. The novel feature of the method is that the data to train the model is constructed in such a way that it carries the information about the dependencies characteristic to the particular regions of the input space (hydrometeorological and flow conditions). These regions correspond to natural clusters of data in input space. This method is referred to as **UN**certainty **E**stimation based on Local **E**rrors and **C**lustering (UNEEC). This abbreviation may also stand for "uncertainty prediction method developed at **UNE**sco-ihe in the framework of a project funded by **EC**".

6.1   Introduction

In Chapter 5, we presented a novel methodology that uses machine learning method to emulate the results of Monte Carlo (MC) simulations. MC is used when it is necessary to study the influence of uncertainty or inputs on the uncertainty of the model output. This is done by sampling the values of the uncertainty entities and running the model for each sample.

However, in practice engineering decisions are often based on a single optimal model run without any uncertainty analysis. Here the model optimality is understood in the following sense: the model is calibrated, so that the model parameters and structure are such that the model error is at minimum. However even such model simulates or predicts the output variable with errors, so its output contains uncertainty. In this chapter we present a novel methodology to estimate the uncertainty of the optimal model output by analyzing historical model residuals errors. This method is referred to as the **UN**certainty **E**stimation based on Local **E**rrors and **C**lustering (UNEEC) method.

We assume that the historical model residuals (errors) between the observed data y and the corresponding model prediction $\hat{y}$ are the adequate quantitative indicators of the gap between the model and the real-world system or process, and they provide valuable information that can be used to assess the model uncertainty. The residuals and their distribution are often functions of the model input, and, possibly, state variables and can be predicted by building a separate model mapping of the input space to the model residuals, or to the distribution of the model error or at least some of its characteristics. In other words, the idea here is to learn the relationship between the input variables and the distribution of the model errors, and to use this information to determine the distribution of the model error when it predicts the output variable (e.g. runoff) in the future.

In this chapter the model residuals are used to quantify the uncertainty. However, typically some strong assumptions about the distribution of such residuals are made:

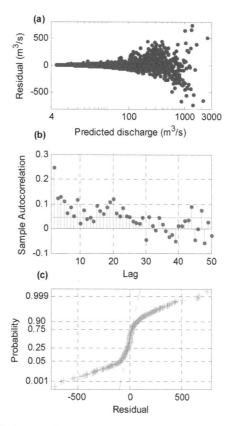

Figure 6.1. Test of residuals analysis. (a) homoscedasticity, (b) independence and (c) normality.

- Model residuals have zero mean and constant variance.

- Model residuals are mutually uncorrelated.

- Model residuals are distributed normally.

Often in practice, most of the above assumptions are always violated (see, e.g., Sorooshian and Dracup, 1980; Xu, 2001). The analysis made for the considered case study (Bagmati catchment, section 7.4) as shown in Figure 6.1 shows that residuals are heteroscedastic, autocorrelated and not normal. This makes it not straightforward to apply existing uncertainty analysis methods, if not impossible. For example, Thyer et al. (2002). applied Box–Cox transformation (Box and Cox, 1964) to normalise hydrological data before performing parameter uncertainty analysis of autoregressive model. Meta-Gaussian model can be applied after transforming model residuals (see section 2.8.5).

Generally the analysis of uncertainty consists of propagating the uncertainty of the input, parameters etc. (which is measured by pdf) through the model by running it for sufficient number of times and deriving the pdf of the model outputs. In this chapter, we present a different approach to analyzing the uncertainty. We focus on residual uncertainty which is defined as the remaining uncertainty of the optimal model. The characteristics of the methodology are:

1. Residuals are used to characterize the uncertainty of the model prediction;

2. No assumptions are made about the distribution of the residuals;

3. The method uses the concept of the model optimality and does not involve any additional runs of the process model;

4. Specialized uncertainty models are built for particular areas of the state space (e.g., hydrometeorological condition). Clustering is used to identify these areas;

5. The uncertainty models are built using machine learning techniques;

6. The method is computationally efficient and can therefore be easily applied to computationally demanding process models; and

7. The method can be applied easily to any existing model no matter whether it is physically based, conceptual or data-driven.

6.2   Definition and sources of model errors

Since the notion of *model errors* is crucial to the UNEEC methodology, it is important to define them. A model error or simply *error* is defined as the mismatch between an observed output value and the corresponding model output or prediction. It is important to distinguish between model residual and model error. The former is the difference between observed output values and the corresponding model outputs in calibration while the latter refers to the difference for prediction/validation.

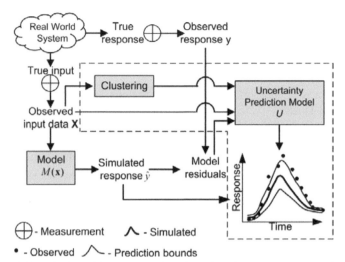

Figure 6.2. Model prediction with the uncertainty bounds using the UNEEC method. The block within the dashed line is the UNEEC method. There is also a linkage between "Model residuals" and "Observed response" to "Clustering".

Deterministic model M (also referred to as "primary model") of a real-world system predicting the system output variable y given the input vector $\mathbf{x}$ ($\mathbf{x} \in X$) can be defined as (see Figure 6.2):

$$\hat{y} = M(\mathbf{x}, \theta) \tag{6.1}$$

where $\hat{y}$ is the model output and θ is a vector of all parameters of the model. Given observed output values of y, the errors of the model predictions are denoted by:

$$\varepsilon = y - \hat{y} \tag{6.2}$$

Equations (6.1) and (6.2) can be summarized by:

$$y = M(\mathbf{x}, \theta) + \varepsilon \tag{6.3}$$

In equation (6.2) ε is the total remaining (or residual) error which incorporates the measurement errors, ε_x, on the input data, measurement errors, ε_y on the output data, the parameter errors, ε_θ, and the model structure errors, ε_s (see section 2.4 for descriptions of the different sources of uncertainties). It can be decomposed as:

$$y = M(\mathbf{x}, \theta) + \varepsilon_s + \varepsilon_\theta + \varepsilon_x + \varepsilon_y \tag{6.4}$$

In most practical cases, however it is difficult to estimate the error components of equation (6.4) unless some very strong assumptions are made. Thus, the different components that contribute to the total model errors are generally treated as a single lumped variable as given in equation (6.3). In this chapter by the model error we understand the total residual error ε which is an aggregate effect of all sources of errors.

6.3   Methodology

The UNEEC method estimates the uncertainty associated with the given model structure M, and the parameter set θ by analysing the historical model residuals ε, which is an aggregate effect of all sources of errors. Thus, the uncertainty estimated with the UNEEC method is valid only for the given model structure and parameter set θ. It does not mean that the model structure and parameter uncertainty are ignored, but it is assumed that the uncertainty associated with the wrong model structure, inaccurate parameter values, and observational errors (if any) are manifested implicitly in the model residuals. This type of uncertainty analysis based on the model residuals is different from the classical uncertainty analysis methods where uncertainty of parameters, input data (presented by pdf) or plausible model structures are propagated to the pdf of the output.

The UNEEC method starts by selecting the single best model structure from the plausible model structures in reproducing the observed behaviour of the system. This ensures that the uncertainty associated with the wrong choice of the model structure is reduced as much as possible. Then it requires the prior identification of an optimal model parameter set, which can be achieved by a calibration procedure aimed at minimizing some error function. This ensures minimizing the uncertainty associated with an inaccurate estimate of parameter values. Observational errors can be reduced by the improved observational techniques and understanding of the characteristics of such errors. There still remains uncertainty which cannot be reduced further because the model error ε cannot be eliminated completely. This uncertainty is defined here in this thesis as *residual uncertainty*. The best we can do is to use this optimal model to predict the output variable of interest and provide the uncertainty estimation based on the analysis of the historical residuals. The aim here is to build a model to estimate the pdf of the model error ε conditioned to the input and/or state variables of the process model.

Since the predictive uncertainty of the model output is more important than the pdf of the model error, the latter is then transferred to the predictive uncertainty by using information on the model predictions (described later in the section). Note that even when the optimal process model is used to produce a deterministic prediction, it does not, however, exclude the possibility of using some combination (ensemble) of "good" (but not optimal) models having the same structure but different in the values of the parameters – which could result from a Monte Carlo exercise. This possibility, however, is not addressed in this study.

The main steps in the UNEEC methodology are:

1. Clustering the input data in order to identify the subsets of more-or-less similar points;

2. Estimating the probability distribution of the model residuals for the subsets identified by clustering;

3. Building the machine learning model of the probability distribution of the model error.

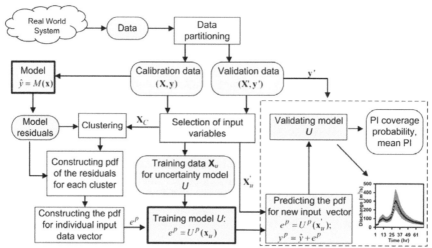

Figure 6.3 The generalized framework of the UNEEC method. The UNEEC method has three steps: clustering, estimation of the probability distribution of the model error, and building model U for probability distribution of the error. Once the model U is trained in the calibration data set (X_u), the model can be used to predict the probability distribution of the model error in a new or unseen data set (e.g., validation data).

These steps are described in the following subsections (see Figure 6.3).

6.3.1 Clustering input data

Clustering of data is an important step of the UNEEC method. Its goal is to partition the data into several natural groups that can be distinguished. By data we understand here the vectors of some variable (input) space, and input space here means not only the input variables of the process model, but also all the relevant state variables which characterize different mechanisms of the modelled process, e.g. runoff generation process. The input data which belong to the same cluster will have similar characteristics and correspond to similar real-life situations. Furthermore, the distributions of the model errors within different clusters have different characteristics. This seems to be a strong assumption of the UNEEC method, which would be reasonable to test before applying it. In hydrological modelling this assumption seems to be quite natural: a hydrological model is often inaccurate in simulating extreme events (high consecutive rainfalls) which can be identified in one group by the process of clustering – resulting in high model residuals (wide error distribution). When data in each cluster belongs to a certain "class" (in this case, a hydrological situation), local error models can be built: they will be more robust and accurate than the global model which is fitted on the whole data.

Before clustering the input data, the most relevant input variables should be selected from the data D. Since clustering is unsupervised learning where the desired output of the system being studied is not known in advance, the selection of the variables in application to the process model is done by incorporating domain (hydrological) knowledge. The data set X_c is the matrix of input data constructed from the data D for

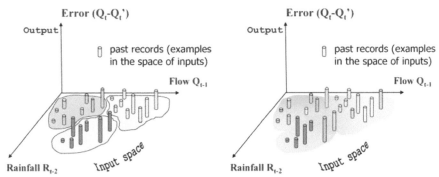

Figure 6.4. Example of clustering the input space of rainfall-runoff model. (a) Hard clustering (e.g. K-means) and (b) soft clustering (e.g. fuzzy C-means clustering).

partitioning into several natural groups (clusters). Additional analysis between model residuals and variables constituting D may be needed to choose the appropriate variables for building X_c from D. Often X_c encompasses the variables of D and additional lagged vectors of some subset of variables of D based on correlation and/or average mutual information (AMI) analysis.

In order to partition the input space X_c, different clustering algorithms (e.g., K-means, fuzzy clustering) can be used. Due to the fuzzy nature of many practical problems, fuzzy clustering has an advantage over K-means clustering by explicitly taking into account the uncertainty associated with the vagueness and imprecision of the uncertainty sources. Figure 6.4 shows the example of clustering in rainfall-runoff modelling. The inputs to the clustering are constructed from the input and output of the rainfall-runoff model.

6.3.2 Estimating probability distribution of model errors

Typically the process model is non-linear and contains many parameters. This will hinder the analytical estimation of the pdf of the model error. Thus the empirical pdf of the model error for each cluster is independently estimated by analysing historical model residuals on the calibration data. In order to avoid a biased estimate of the pdf or its quantiles of the model error, it is important to check if there is any over-fitting by the process model on the calibration data. It is also possible to use leave-one-out cross-validation (Cawley et al., 2004) to overcome the bias estimate of the quantiles if the computational burden of running the process model is not prohibitive. However, such a cross-validation technique may be impractical in hydrological modelling because of the computational load resulting from training multiple models. Another solution is to use a separate calibration sample data set that has not been used to calibrate the model, provided enough data are available. Note that when dealing with limited calibration data, the empirical distribution might be a very poor approximation of the theoretical distribution, so the reliability of such a method depends on the availability of data.

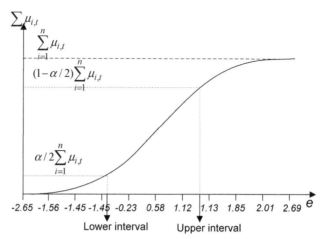

Figure 6.5. Example of cumulative distribution function of the model error weighted by fuzzy membership function. In this example "lower interval" corresponds to the $\alpha/2 \cdot 100$ % error quantile and "upper interval" corresponds to the $(1-\alpha/2) \cdot 100$ % error quantile, where α is significance level $(0 < \alpha < 1)$ and $1-\alpha$ is corresponding confidence level.

Since the pdf of the model error is estimated for each cluster, it depends on the clustering method used. For example, in the case of K-means clustering where each instance of data belongs to only one cluster, the quantiles are taken from the empirical error distribution for each cluster independently. However, in the case of the fuzzy clustering method (FCM) where each instance belongs to more than one cluster, and is associated with several membership functions, the computation of the quantiles should take this into account. The pth $[0, 1]$ quantile of the model error for cluster i is given by:

$$ec_i^p = \varepsilon_t \qquad t : \sum_{k=1}^{t} \mu_{i,k} < p \sum_{t=1}^{n} \mu_{i,t} \qquad (6.5)$$

where t is the maximum integer value running from unity that satisfies the above inequality, ε_t is the residual associated with the tth data (data are sorted with respect to the associated residual), and $\mu_{i,t}$ is the membership function of the tth data to cluster i. This is not the only way of calculating quantiles for fuzzy clusters. An alternative would be to use the threshold of the membership degree in selecting the points to be included in sorting for each cluster.

Figure 6.5 shows the cumulative distribution function of the model errors weighted by corresponding fuzzy membership function, from which any desired pth quantile can be estimated by linear interpolation. The figure also shows the example of computing $(1-\alpha) \cdot 100$ % prediction intervals of the model error by estimating two quantiles, i.e., $\alpha/2 \cdot 100$ % error quantile (often termed lower intervals) and $(1-\alpha/2) \cdot 100$ % error quantile (upper intervals).

6.3.3   Building model for probability distribution of model errors

Having the pdf or some quantiles of the process model error for each individual cluster $\{ec_i^P\}_{i=1}^c$ and outputs of the clustering i.e. partition matrix and cluster centres, the next step is to approximate the pdf or quantile of the model error for the input data vector $\mathbf{x}_t$. The machine learning model which is built to predict the pdf of the model error for the new data input, is referred to as an "uncertainty model U" and can be built using several approaches (Shrestha and Solomatine, 2008) as presented below.

Using eager learning classifiers

The goal here is to build a classifier as an uncertainty model U given the input data matrix X, results of clustering analysis $\{P, V\}$, and the quantile of the model error for each cluster $\{ec_i^P\}_{i=1}^c$, where P is the partition matrix and V is the matrix of clustering centres (see section 3.8). First, the classifier is trained on the input data matrix X_u and the cluster labels $\{C_i\}_{i=1}^c$ as class values (i.e. this classifier uses so-called eager learning). Second, the trained classifier is used to classify the new data input $\mathbf{x}_t$ to one of the cluster. Last, the quantile of the model error for the new data input $\mathbf{x}_t$ is approximated by taking the quantile of the corresponding clusters.

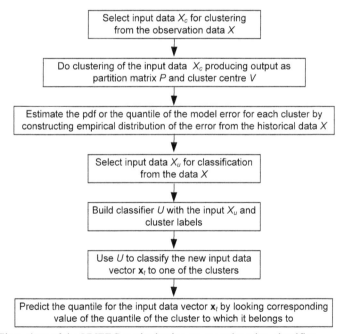

Figure 6.6. Flow chart of the UNEEC method using an eager learning classifier.

The choice of variables for input data matrix X_u may require some analysis to ensure the accuracy of the classifier. This method is applicable for non-overlapping clustering (such as, for instance, K-means clustering). The method ignores the variation of the error distribution (or quantiles) inside the cluster, hence the input data which belongs to the same cluster has the same value of the quantile of the model error distribution. The disadvantage of the method is that the transition between the quantile estimated across the input data is not smooth. The flow chart to estimate the quantile of the model error by this method is given in Figure 6.6.

Using a committee of instance-based (lazy) learning classifiers

In instance-based learning methods, instead of building a classifier, a distance function $d(\mathbf{v}_i, \mathbf{x}_t)$ between the centres of the cluster C_i (where $\mathbf{v}_i$ is the vector of centre of the cluster C_i) and the new input data $\mathbf{x}_t$ is used to classify $\mathbf{x}_t$ to the particular cluster. The new data $\mathbf{x}_t$ will be assigned to the cluster C_i such that a distance function $d(\mathbf{v}_i, \mathbf{x}_t')$ is minimum. Mathematically it is expressed as:

$$\mathbf{x}_t \in C_i : \min(d(\mathbf{v}_i, \mathbf{x}_t)) \tag{6.6}$$

Another variant of instance based learning for the estimation of the quantiles of the model error using the so called *fuzzy committee* approach is shown in Figure 6.7. Instead of assigning an input vector to a particular cluster, fuzzy membership weights $w_{i,t}$ are computed using the distance functions $d(\mathbf{v}_i, \mathbf{x}_t)$ between input data $\mathbf{x}_t$ and the cluster centres $\{\mathbf{v}_i\}_{i=1}^c$. These fuzzy membership weights $\mu_{i,t}$ reflect the proximities of the input data to the cluster centres. There are many different ways of defining a distance function, and it is hard to find rational grounds for choosing any one in particular. Typically, the relation between weight and distance functions takes the form of

$$w_{i,t} = d(\mathbf{v}_i, \mathbf{x}_t)^{-2/m} \tag{6.7}$$

where m is the smoothing exponential coefficient, $d(\mathbf{v}_i, \mathbf{x}_t)$ is the ordinary Euclidean distance between cluster centres $\mathbf{v}_i$ and the input data vector $\mathbf{x}_t$. The smaller value of m (≈ 0.25) gives relatively higher weight to the least distance while the higher value of it (≈ 100) gives almost equal weights.

Once the fuzzy weights are computed, the quantiles of the model error for the new input data $\mathbf{x}_t$ can be computed using the fuzzy committee approach as:

$$e_t^p = \sum_{i=1}^c w_{i,t}^{2/m} \, ec_i^p \,/ \sum_{i=1}^c w_{i,t}^{2/m} \tag{6.8}$$

where e_t^p is the pth quantile of the pdf of the error. It should be noted that equation (6.8) is indeed a soft combination of the quantiles of each cluster depending upon the membership function values. This formulation has an additional advantage of

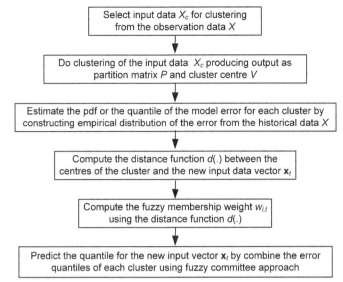

Figure 6.7. Flow chart of the UNEEC method using fuzzy committee of instance based classifiers.

smoothing the quantiles across the input data. The smoothing can be increased with a higher value of m (e.g., $m = 100$ gives almost equal weights for all clusters).

Using regression methods

In the presented methods, first the pdf (or its quantiles) of the model error is estimated for the individual input data vector where the information about the model residuals is known (e.g., in the calibration data). Since the empirical pdf of the model error for the clusters are already computed following the method described in the subsection 6.3.2, the input data being the member of the clusters share this information of distribution.

(It is worth mentioning that the estimation of quantiles for the individual input data vector depends on the types of clustering techniques employed. For example, in K-means clustering the input data share the same information of the pdf of the error for a particular cluster, thus ignoring the variation of the error distribution inside the cluster. However, there are other possibilities such as using a distance function (distance between the centres of the cluster to the input vector, e.g., equation (6.7)) as a weight to vary the error distribution).

In the case of fuzzy clustering the fuzzy committee approach as described in the section on instance-based learning methods (see equation (6.8)), is used to compute the quantiles for each individual input data vector. Once the quantiles of the pdf of the model error for each example in the training data are obtained, the machine learning model U (that estimates the underlying functional relationships between the input vector x_u and the computed quantiles) is constructed:

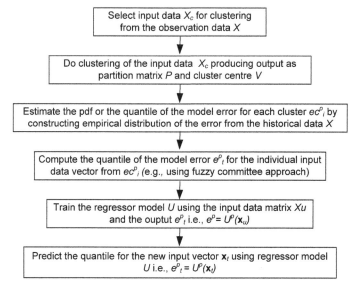

Figure 6.8. Flow chart of the UNEEC method using regression method.

$$e^P = U^P(\mathbf{x}_u; \; \theta^P)$$
(6.9)

where θ^P is the parameters vector of the model U^P for the pth quantile. Note that the calibration data set for the model U is $(X_u, \mathbf{e}^P)$, where X_u is the input data constructed from X described below, and $\mathbf{e}^P$ is a vector of pth quantiles. Thus the model U, after being trained on input data X_u, encapsulates the pdf of the model error and maps the input $\mathbf{x}_u$ to the pdf or the quantiles of the process model error.

The model U can take any form, from linear to non-linear regression models, such as neural networks. The choice of the model depends on many factors including the complexity of the problem to be handled and the availability of data etc. Once the model U is trained on the calibration data X_u, it can be employed to estimate the quantiles or the pdf of the model error for the new data input (e.g., new hydrometrological condition in rainfall-runoff modelling).

6.4   Computation of predictive uncertainty of model output

In the previous sections 6.3.1-6.3.3, we proposed a method to predict the pdf (or its quantile) of the process model error. As previously mentioned, the predicted quantiles of the pdf of the model error should be transferred to a more meaningful and understandable entity – the predictive uncertainty of the model output. The quantile of the predictive uncertainty of the tth model output can be estimated as:

$$(6.10)$$

$$y_t^p = \hat{y}_t + e_t^p$$

where $\hat{y}_t$ is the tth output simulated by the process model, y_t^p is the pth quantile of the tth model output, and e_t^p is the pth quantile of the model error corresponding to the tth model output. One can see that equation (6.10) is the reformulation of equation (6.2).

Often uncertainty of the model output is represented by the prediction intervals. $\cdot(1 - \alpha)\cdot100\%$ prediction intervals of the model output are estimated by computing the $\alpha/2\cdot100\%$ and $(1-\alpha/2)\cdot100\%$ quantiles of the model output, where $(1 - \alpha)\cdot100\%$ is the confidence level and α is the significance level used to compute the confidence level. For example, α of 0.05 indicates a 95 percent confidence level. These two quantiles $\alpha/2\cdot100\%$ and $(1-\alpha/2)\cdot100\%$ are termed as lower and upper prediction limits of the $(1 - \alpha)\cdot100\%$ prediction intervals, respectively and estimated by equation (6.10) as follows:

$$PL_t^L(\alpha) = y_t^{\alpha/2} = \hat{y}_t + e_t^{\alpha/2}$$
$$PL_t^U(\alpha) = y_t^{(1-\alpha/2)} = \hat{y}_t + e_t^{(1-\alpha/2)}$$

$$(6.11)$$

where $PL_t^L(\alpha)$ and $PL_t^U(\alpha)$ are the lower and upper prediction limits corresponding to α significance level (i.e., $1-\alpha$ confidence level). $e_t^{\alpha/2}$ and $e_t^{(1-\alpha/2)}$ are two error quantiles estimated by building two independent models $U^{\alpha/2}$ and $U^{(1-\alpha/2)}$ following equation (6.8) as

$$e_t^{\alpha/2} = U^{\alpha/2}(\mathbf{x}_u; \ \theta^{\alpha/2})$$
$$e_t^{(1-\alpha/2)} = U^{(1-\alpha/2)}(\mathbf{x}_u; \ \theta^{(1-\alpha/2)})$$

$$(6.12)$$

For example, in order to estimate the 90% prediction interval of the model output, it is necessary to build two models, U^5 and U^{95} that will predict the 5% and 95% quantiles of the model errors, respectively. It is worthwhile to mention that the input variables used to build these two independent models might be different, and are discussed in section 6.5.

6.5   Selection of input variables

The selection of appropriate and relevant input variables for machine learning models used in the UNEEC method is done using the similar approach described in the MLUE method (see section 4.6) and is not described here.

6.6   Validation of UNEEC method

The UNEEC method is validated by: (i) measuring the predictive capability of the uncertainty model U (e.g., using root mean squared error); (ii) measuring the statistics of the uncertainty estimation; and (iii) visualizing plots of the prediction intervals with the observed hydrograph.

Validation of model U, however, is not straightforward due to the following. A validation procedure typically assumes that for a new situation (e.g., in rainfall-runoff model, new values of rainfall, runoff, etc.) it is possible to compare the output variable value calculated by the model (say, the 5% quantile calculated by the model U^5) to the corresponding "measured" value. However, it is impossible to measure the value of a quantile (in this case a 5% quantile) whereas it is possible to validate the performance of model M since it is possible to measure the runoff. One may argue that if there is a large validation data set available, it is possible to generate the quantiles in the validation data set either (i) directly from the whole set; or (ii) from the clusters found in the validation set using the same procedure adopted when building the model U. However, in option (i) the comparison would not be fair since the model U uses different (local, cluster-based) models of uncertainty, and in option (ii) the clusters generated will be different from those generated during building of U and hence the comparison would not be proper either.

Shrestha et al. (2006) proposed a partial solution to validate the model U. This is described here. The calibration data X_u was divided into two parts: *training data*, which constitutes the major part of the calibration data (in this study, 67%), were selected randomly from the calibration pool without replacement. The training data are used to train the model U. The remaining data from the pool is the *cross-validation* (or test) data set which is used to perform "intermediate" tests of the model U with the purpose to select its structure, parameters and the input variables. Random (or close to random) selection ensures statistical similarity of these data sets. Once the model U is tested on the cross-validation (test) data set, the model U with the best structure can be re-trained on the full calibration data so as to increase its accuracy.

Even though the quantiles of the model error and, consequently, the quantiles of the model output are not observable in the validation data set, it is interesting to know if the distribution of the observed data fits the distribution predicted by model U. When predicting only two quantiles (or prediction interval, PI), this problem is equivalent to counting how many of the observed values are inside the prediction interval. In this case, validation of the UNEEC method can be done by evaluating two statistics: the prediction interval coverage probability (PICP) and the mean prediction interval (MPI). These uncertainty statistics is described in section 4.7. In principle, these statistics can also be computed individually for each cluster by training a classifier that would be able to attribute the new input vectors to one of the clusters or by computing the distance between the new input vectors and the cluster centroid vectors.

6.7 Limitations and possible extensions of method

Machine learning methods, which constitute the core of the UNEEC method, have particular limitations. These limitations have been already discussed in section 4.8 in relation to the MLUE method and most of them are also valid in relation to the UNEEC method.

UNEEC method allows for useful extensions. As mentioned above, this method relies on the concept of model optimality instead of equifinality. If the assumption of the existence of the single "best" model is not valid, then all of the models that are considered "good" should be considered, as it is done when the concept of equifinality is adopted. This can be achieved by combining such models in an ensemble, or by generating the meta models of uncertainty for each possible combination of the model structure and parameter set, or even involving the uncertainty associated with the input data. Consequently, instead of having a single set of uncertainty bounds for each forecast, there will be a set of such bounds generated. However, the use of such several uncertainty bounds in the decision making process would be really challenging.

Chapter 7
Application of Machine Learning Method to Predict Residual Uncertainty

This chapter presents the application of UNEEC (machine learning method to predict residual uncertainty) method to estimate uncertainty in the model prediction for the four different case studies. In first one, synthetic data set (where the exact solution is known) is used to demonstrate the method's capability. The second case study is a typical river flow forecasting problem of the Sieve catchment in Italy. In this study the river flow forecasts are made by several data-driven techniques such as artificial neural networks, model trees, and locally weighted regression. The third case study considers a conceptual rainfall-runoff model of the Brue catchment in United Kingdom. The Hydrologiska Byråns Vattenbalansavdelning (HBV) model is used to simulate hourly river flows data. In the fourth case study, we apply the UNEEC method to a rainfall-runoff model of the Bagmati catchment in Nepal. The HBV model is used to simulate daily river flows data. In all these case studies, the results are compared with other existing uncertainty methods including the generalised likelihood uncertainty estimation (GLUE), the meta-Gaussian, and the quantile regression methods.

7.1   Application 1: Synthetic data set[2]

We apply the UNEEC method to estimate the uncertainty of model prediction for a synthetic data set. Synthetic data set is generated by simple linear regression models where there exists an analytical solution for uncertainty estimation.

---

[2] Based on: Shrestha, D.L. and Solomatine, D. (2006). Machine learning approaches for estimation of prediction interval for the model output. *Neural Networks*, 19(2), pp. 225-235.

7.1.1 Linear regression model of synthetic data set

The synthetic data is generated using bivariate input variables $\mathbf{x}$ uniformly distributed between [0, 1]. The true targets y_t are given by:

$$y_t = f_t(\mathbf{x}) = b_1 x_1 + b_2 x_2 \tag{7.1}$$

where b_1 and b_2 are linear regression coefficients and arbitrary values of these coefficients: $b_1 = 12$ and $b_2 = 7$ are used in the experiments. Since noise is inherent to any real data set, we also add additive noise ε to y_t to obtain a new target y, which is estimated by the function f_f as:

$$y = y_t + \varepsilon = f_f(\mathbf{x}) \tag{7.2}$$

The noise ε has a Gaussian distribution $N(0, \sigma_t/\text{SNR})$. SNR is the signal-to-noise ratio defined by:

$$\text{SNR} = \sigma_t / \sigma_\varepsilon \tag{7.3}$$

where σ_t and σ_ε are the standard deviations of the target y and noise ε, respectively.

Given training data set $\{\mathbf{x}_i, y_i\}$, $i = 1, \ldots, n$, where n is the number of training data, the problem is to find the coefficients b_1 and b_2 using the least squares method. Once the linear regression model is built, the regression model is used to predict the new data vector while UNEEC method is used to estimate the uncertainty of the model prediction.

7.1.2 Prediction interval for linear regression

In regression problems, the task is to estimate an unknown function $f(\mathbf{x}; \theta)$ given a set of input-target pairs $D = \{\mathbf{x}_i, y_i\}$, $i = 1, \ldots, n$, where θ is the true values of the set of parameters of the regression model. The function $f(\mathbf{x}; \theta)$ is related with the true target y by

$$y_i = f(\mathbf{x}_i; \theta) + e_i \tag{7.4}$$

where e_i is the model error and is assumed to be independently and identically (iid) distributed with variance σ^2 and the distribution has the form $N(0, \sigma^2)$. The least-square estimate of parameters θ is $\hat{\theta}$, which is obtained by minimizing the cost function:

$$C(\theta) = \sum_{i=1}^{n} [y_i - \hat{y}_i]^2 \tag{7.5}$$

$$\hat{y}_i = f(x_i; \hat{\theta}) \tag{7.6}$$

where $\hat{y}_i$ is the model output. The model output may not necessarily match the target y due to various reasons including the presence of noise in the data, limited number of data points, non-linear relationship between the dependent and independent variables, and the errors in estimating the parameters. In order to simplify the formulation, it is assumed that the regression model is linear and univariate; so equation (7.6) can be rewritten as:

$$\hat{y}_i = \hat{a}x_i + \hat{b} \tag{7.7}$$

where $\hat{a}$ and $\hat{b}$ are estimated parameters of the linear regression.

Most of the methods to construct $(1-\alpha) \cdot 100\%$ prediction limit for the model output typically assume the error has a Gaussian distribution with zero mean and σ standard deviation. The prediction limits are given by:

$$PL^U = \hat{y} + z_{\alpha/2}\sigma \tag{7.8}$$
$$PL^L = \hat{y} - z_{\alpha/2}\sigma$$

where PL^U and PL^L are the upper and lower prediction limits, respectively, $z_{\alpha/2}$ is the value of the standard normal variate with cumulative probability level of $\alpha/2$. Since the prediction limits in equation (7.8) are symmetric about $\hat{y}_i$, it is assumed that the prediction is unbiased. Consequently, the model error variance σ^2 equals the *mean squared error* (MSE), often called the *prediction mean squared error* (PMSE) (Chatfield, 2000). However, σ^2 is not known in practice and is estimated from the data. An unbiased estimate of σ^2 with $n - p$ degrees of freedom, denoted by s^2, is given by the formula:

$$s^2 = SSE / (n\text{-}2) = \frac{1}{n-2} \sum_{i=1}^{n} (y_i - \hat{y}_i)^2 \tag{7.9}$$

where p is number of parameters in the model and SSE is the *sum squared error*.

If the error variance s^2 is not constant in the output space, then equation (7.9) should be modified as follows (Harnett and Murphy, 1980; Wonnacott and Wonnacott, 1990):

$$s_{y_i}^2 = s^2(1 + 1/n + (x_i - \overline{x})^2 /(n-1)s_x^2) \qquad (7.10)$$

where s_x^2 and $\overline{x}$ are the sample standard deviation and mean, respectively. It can be seen that the error variance for the output y_i is always larger than s^2 and it is dependent on how far x_i is away from $\overline{x}$. For multivariate linear regression equation (7.10) can be modified as:

$$s_{y_i}^2 = s^2(1 + x_i^T (X^T X)^{-1} x_i) \qquad (7.11)$$

where X is a matrix of input space, appended to a column of 1's as the leftmost column and x_i is i^{th} row of matrix X. In this thesis, this method will be referred to as *analytical approach* method.

7.1.3 Experimental setup

Total of 1500 bivariate data are generated using equation (7.2). We vary the noise level by taking the following SNR values: 1, 3, 5 and 7. Two-thirds of data set is selected randomly to constitute the training data set, and the rest constitutes the test or verification data set.

For a given data set, the prediction or primary model f_f was built using the training data set. The model's objective of which is to estimate the target t as closely as possible. Then the trained model was used to predict the targets for the new input data set. Having the model outputs for the test data set, the UNEEC method was applied to estimate the PI on the test data set as follows. The Fuzzy C-means clustering technique was first employed to construct the prediction interval (PI) for each cluster in the training data set and then to construct the PI for each instance in the training data set. Note that the input to the uncertainty model U for the UNEEC method may constitute part or all of input variables, which are used in the prediction model f_f. The targets for the uncertainty model U are the upper and lower PIs.

Table 7.1. Summary of input data and regression models used in the experiments.

Experiment	SNR	Target variable	Input variables for primary model, clustering and UNEEC method	Model Primary	UNEEC
S1	1	y	x_1, x_2	LR	LR
S3	3	y	x_1, x_2	LR	LR
S4	4	y	x_1, x_2	LR	LR
S7	7	y	x_1, x_2	LR	LR

The PIs are constructed for 95% confidence level unless specified. The performance of the UNEEC model is assessed by the prediction interval coverage probability (PICP) and the mean prediction interval (MPI). Four experiments are considered as shown in Table 7.1.

7.1.4 Results and discussions

The two performance indicators: PICP and MPI are calculated for the UNEEC and the analytical approach (see Table 7.2), and are shown in Figure 7.1. It is worthwhile to mention that the linear regression model is used as an uncertainty model in the UNEEC method. It can be seen that the UNEEC method appears to perform better for data with high noise if compared to the analytical approach. For lower noise the PICPs statistics of the UNEEC method are close to desired degree of confidence level (i.e. in this case 95%). However, the PICP for the analytical approach deviates slightly from the desired 95% confidence level. For all noise levels, the MPI values for the analytical approach are wider than those for the UNEEC method. Note that the uncertainty increases as the noise increases.

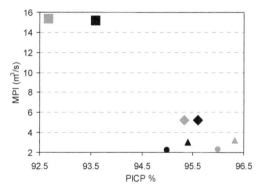

Figure 7.1. Comparison of performance. Black blocks refer to the UNEEC method, the grey blocks—to the analytical approach. The size of marker represents the noise level (i.e., bigger marker – higher noise).

Table 7.2. Comparison of results on test data set for linear regression models

Experiment	Prediction error (m³/s)	Analytical method		UNEEC method	
		PIPC (%)	MPI (m³/s)	PIPC (%)	MPI (m³/s)
S1	4.01	92.67	15.33	93.60	15.17
S3	1.34	94.80	5.24	95.60	5.21
S4	0.77	96.60	3.22	95.40	3.05
S7	0.58	95.40	2.28	95.00	2.26

7.1.5 Conclusions

In this study, we apply the UNEEC method to a synthetic data set generated by simple linear regression model where there exists an analytical solution for uncertainty estimation. We add noises to the output variable of synthetic data set. The results obtained with the synthetic data set show that the UNEEC method gives the consistently narrower uncertainty bounds than those of the analytical approach. The uncertainty obviously increases when the noise is added to the data. This simple application demonstrates that the UNEEC method is as good as the analytical approach and method can be used for real life applications.

7.2   Application 2: Sieve River catchment[3]

This application presents the UNEEC method to a typical river flow forecasting problem. Flow forecasts with several lead times are made by data-driven modelling techniques, namely artificial neural networks (ANN), model trees (MT), locally weighted regression (LWR). Then we apply the UNEEC method to estimate the uncertainty in the flow forecasting for a medium sized catchment, namely the Sieve catchment in Italy.

7.2.1   Flow forecasting model of the Sieve River catchment

Sieve is a tributary of the Arno River (see Figure 7.2) and located on the Apennines Mountains in Tuscany, North-Central Italy. The Sieve catchment consists mostly of hills, forests and mountainous areas except in the valley with an average elevation of 470 m above sea level. The climate of the catchment is temperate and humid. The catchment has an area of 822 Km^2. 3 months of hourly runoff discharge, precipitation and potential evapotranspiration data were available (December 1959 to February 1960), which represent various types of hydrological conditions and flows range from low to very high.

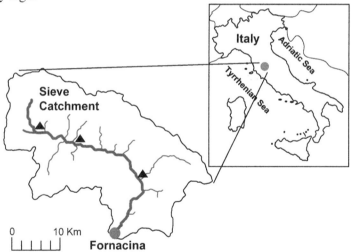

Figure 7.2. Sieve catchment. Triangles show the rainfall stations; the flow is measured at the Fornacina gauge station.

The discharge data were available at the closure section of Fornacina. The spatial average of hourly rainfall from 11 rain gauge stations was calculated by Thiessen polygon method and hourly evapotranspiration data were calculated using radiation

[3]   Based on: Shrestha, D.L. and Solomatine, D. (2006). Machine learning approaches for estimation of prediction interval for the model output. *Neural Networks*, 19(2), pp. 225-235.

method. The Arno catchment that includes the Sieve catchment has been extensively studied in various modelling exercises (Todini, 1996; Solomatine and Dulal, 2003; Solomatine et al., 2008).

We consider three forecasting models corresponding to 1, 3 and 6 hours lead time denoted by SieveQ1, SieveQ3 and SieveQ6, respectively. For each model, several machine learning techniques, namely ANN, MT, LWR and linear regression are built. The selection of the input variables for the forecasting models is based on the methods discussed in section 3.8.3. Visual joint inspection of a number of peak rainfall events and the resulting hydrographs shows that on average the time lag between several peak rainfalls and runoffs is between 5 and 7 hours. Additional analysis of lags is performed using the average mutual information (AMI) and cross correlation analysis of rainfall and runoff. The cross correlation between the rainfall and runoff increases with the lag, reaches a maximum of 0.75 when the lag is 6 hours, and then starts to decreases (Figure 7.3). These results are consistent with the AMI analysis.

Besides rainfall, previous flows can also be used as input variables in the model, since they also have a high correlation with the future flows. The choice of the appropriate number of previous flows is based on analysing their autocorrelation. The autocorrelation coefficient is close to unity for a 1 hour lag and decreases as the lag increases. Based on these analyses, forecasting models of the Sieve catchment are formulated as below:

$$Q_{t+1} = f_f(RE_t, RE_{t-1}, RE_{t-2}, RE_{t-3}, RE_{t-4}, RE_{t-5}, Q_t, Q_{t-1}, Q_{t-2})$$
$$Q_{t+3} = f_f(RE_t, RE_{t-1}, RE_{t-2}, RE_{t-3}, Q_t, Q_{t-1})$$
$$Q_{t+6} = f_f(RE_t, Q_t)$$

(7.12)

where, $Q_{t+1}, Q_{t+3}, Q_{t+6}$ are flows at 1, 3, and 6 hours lead time.

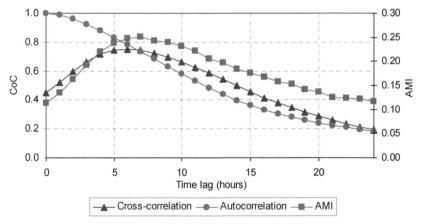

Figure 7.3. Correlation and average mutual information (AMI) analysis between the forecast variable Q_{t+1} and basin average rainfall and autocorrelation of Q_{t+1}.

Table 7.3. Statistical properties of the runoff data sets

Statistical properties	Available data	Training set	Verification set
Period (yyyy/mm/dd hh:mm)	1959/12/01 07:00 - 1960/02/28 00:00	1959/12/13 19:00 - 1960/02/28 00:00	1959/12/01 07:00 - 1959/12/13 18:00
Number of data	2154	1854	300
Average (m^3/s)	54.9	53.1	66.8
Minimum (m^3/s)	10.7	10.7	17.6
Maximum (m^3/s)	752.6	752.6	299.9
Standard deviation (m^3/s)	70.2	73.0	48.7

Splitting of the available data set into training and testing is done based on the research of Solomatine and Dulal (2003) and Solomatine et al. (2008) and is not discussed here. Records between 1959/12/13 19:00 and 1960/02/28 00:00 are used to form the training data set, and the first 300 records (1959/12/01 07:00 to 1959/12/13 18:00) are used as the test data. The some statistical properties of runoff data sets are presented in Table 7.3.

7.2.2 Experimental setup

Several linear and non-linear models are built for the river flow forecasting of the Sieve River catchment. As in the case of using the synthetic data set, first the primary forecasting model f_f is built using the training data set. Then the UNEEC method is applied to estimate the forecast uncertainty. In principle, the structure or even the class of the primary model and the UNEEC model U can be different (for example we may use neural networks for the primary model and linear regression for the UNEEC model). For the purpose of comparison to other methods, we employ the same class of regression model. Table 7.4 presents the summary of the input data and regression models used in the experiments.

Table 7.4. Summary of input data and regression models used in the experiments.

Experiment	Target variable	Input variables for primary model, clustering and UNEEC method	Model Primary	Model UNEEC
SieveQ1Bivar	Q_{t+1}	RE_{t-5}, Q_t	LR	LR
SieveQ3Bivar	Q_{t+3}	RE_{t-2}, Q_t	LR	LR
SieveQ1	Q_{t+1}	RE_t, RE_{t-1}, RE_{t-2}, RE_{t-3}, RE_{t-4}, RE_{t-5}, Q_t, Q_{t-1}, Q_{t-2}	LR, LWR, ANN, MT	LR, LWR, ANN, MT
SieveQ3	Q_{t+3}	RE_t, RE_{t-1}, RE_{t-2}, RE_{t-3}, Q_t, Q_{t-1},	LR, LWR, ANN, MT	LR, LWR, ANN, MT
SieveQ6	Q_{t+6}	RE_t, Q_t	LR, LWR, ANN, MT	LR, LWR, ANN, MT

Note: SieveQ1Bivar and SieveQ3Bivar represent the experiments with bivariate inputs to forecasting models for 1 hour and 3 hours lead time, respectively. LR – linear regression.

Bivariate linear regression model is built using only the two most influencing input variables (i.e., variables with the highest correlation and AMI with the output). The input variables are extended according to equation (7.12). To investigate the effect of models' complexity on the estimated uncertainty bounds, experiments are also conducted using ANN, MT and LWR as both primary and uncertainty models.

7.2.3 Results and discussions

Analysis of clustering

Figure 7.4 shows the Xie-Beni separation index S used to determine the optimal number of clusters. This number corresponds to the minimum value of the S index. The optimal number of clusters is between 4 and 6. We also validate the optimal number of clusters by estimating the PIs using different numbers of clusters. The results are presented in Figure 7.5. It is observed that the percentage of the observed discharge falling inside the computed 95% PIs (i.e. PICP) is close the desired degree of confidence level (i.e. 95%) at 6 number of clusters. However, the number of clusters corresponding to the minimum MPI is 4. In order to balance PICP and MPI, the number of clusters chosen for all the experiments was 5. Note that all the experiments were performed with the exponential exponent m equal to 2 in fuzzy clustering. As the value of m approaches unity, clustering is close to the exclusive clustering (e.g. *K-means*), and as the value of m increases, the degree of fuzziness increases.

Figure 7.6 shows the clustering of input examples in the Sieve catchment for a 1 hour ahead prediction of the runoff (SieveQ1 data set). The results show that the input examples with very high runoff have maximum fuzzy membership function values to Cluster 1 (denoted by C1). Whereas the input examples with very low values of runoff have maximum fuzzy membership function values to Cluster C5. Table 7.5 shows the computed PIs for the clusters for the SieveQ1 data set using the 95% degree confidence level (i.e. $\alpha = 0.05$). The results from the prediction model (in this case MT) are biased and the model underestimates the target value Q_{t+1} in the SieveQ1 data set. Bias in the model predictions arises for many reasons including the skewed distribution of the

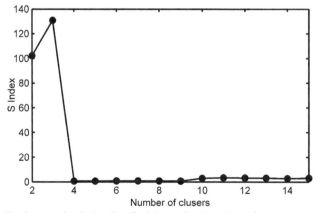

Figure 7.4. Xie-Beni separation index S to find the optimal number of clusters.

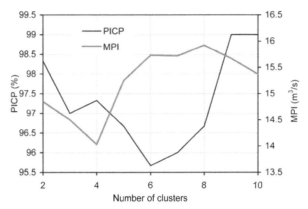

Figure 7.5. Sensitivity of PICP and MPI with different numbers of clusters for the SieveQ1 data set. M5 model trees are employed as the primary and the UNEEC model.

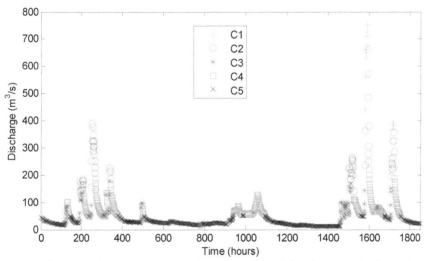

Figure 7.6. Clustering of the input examples in the SieveQ1 training data set using fuzzy C-means clustering. The figure shows the maximum memberships of the input examples to the clusters.

training data, the selection of non-optimal values of the model parameters, the optimisation techniques etc. For example, the hydrological river flow data is often fitted with a lognormal probability distribution (Kottegoda, 1980). Consequently, the computed PIs are asymmetrical and the absolute values of the upper PIs are greater than those of the lower PIs.

In order to select the most important influencing variables of the model, including unsupervised learning such as clustering; several approaches have been reported in the literature. We use the linear correlation analysis in this study. Figure 7.7 shows the correlation analysis of the computed upper PIs in the training data set and the input data for different values of the lag time. It can be observed that the upper PIs have the

Table 7.5. Cluster centres and computed prediction intervals for each cluster in the SieveQ1 data set.

Cluster ID	Cluster centre			Prediction Interval (m³/s)	
	RE_{t-5} (mm/hour)	$Q_t (m^3/s)$	$Q_{t-1} (m^3/s)$	Lower	Upper
C1	5.81	656.99	644.19	-106.99	134.84
C2	1.96	260.01	257.53	-52.50	66.28
C3	0.56	119.38	119.75	-18.05	24.57
C4	0.08	59.74	60.38	-6.08	6.7
C5	0.04	23.00	23.05	-1.01	1.42

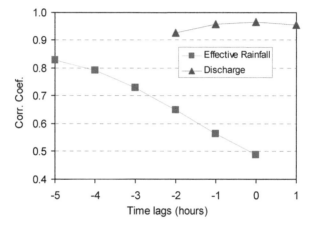

Figure 7.7. Correlation between computed upper prediction interval with input data effective rainfall, RE_t and discharge, Q_t for different values of time lags. Upper prediction interval has highest correlation with RE_{t-5} (time lag is 5 hr). Similarly upper prediction interval has highest correlation with Q_t.

highest correlation with RE_{t-5} (lag time is 5 hr) and that they decrease slowly as the lag time decreases. Similarly, runoff of the present time (Q_t) contains much more information than that of previous values (say up to a 2 hour lag) of the runoff to estimate upper PIs of the 1 hour ahead runoff (Q_{t+1}), because the upper PIs have the highest correlation with Q_t. Similar results are also obtained for the lower PIs.

7.2.4 Uncertainty estimation

The performance of the UNEEC method is compared to that of the analytical approach in the case of experiments with the bivariate input variables (see Table 7.6). It is observed that the UNEEC method shows superior performance for both time leads of the forecasts with respect to PICP and MPI.

Table 7.6. Results on the test data set for hydrological data set using M5 model trees as primary and uncertainty model in the UNEEC method.

Experiment	Prediction error (m³/s)	Analytical or uniform interval method		UNEEC method	
		PIPC (%)	MPI (m³/s)	PIPC (%)	MPI (m³/s)
SieveQ1Bivar	6.13	98.00	40.98	99.93	25.61
SieveQ3Bivar	14.18	98.00	86.90	98.00	61.76
SieveQ1	3.61	96.67[a]	15.25[a]	91.33	11.80
SieveQ3	13.67	95.67[a]	43.27[a]	89.33	40.58
SieveQ6	22.89	97.67[a]	96.34[a]	91.33	81.6

Note: [a] – uniform interval method

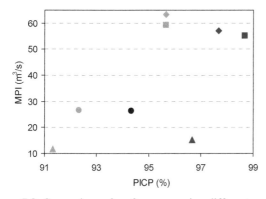

Figure 7.8. Comparison of performance using different machine learning techniques for SieveQ1 data set. Black blocks refer to the UNEEC methods, the grey blocks – to the analytical approach. Rhombus, square, circular and triangle shapes represent linear regression, locally weighted regression, artificial neural network and model tree respectively.

A comparison of the performance using various data-driven models is shown in Figure 7.8. We employ multiple linear regressions, LWR, ANN and MT to predict runoff 1, 3 and 6 hour ahead. These data-driven models are also used as uncertainty models to estimate the PIs. The results show that the performance of MT is better than that of the other models. The performance of the linear regression and LWR models are comparable.

We compare the results with the *uniform interval method* (UIM) that constructs a single PI from the empirical distribution of errors on the whole training data and is applied uniformly to the test data set. The results are presented in Table 7.6. The UIM method gives a constant width of the uncertainty bounds and is used here as bench mark to compare the results with UNEEC method. The UNEEC method performs consistently better than the UIM as PICPs of UNEEC are closer to the desired confidence level of 95%.

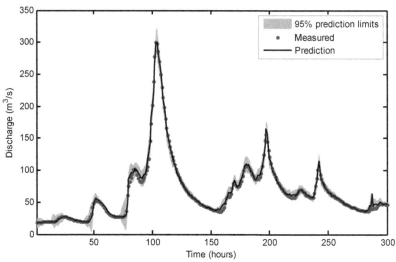

Figure 7.9. Computed prediction limits for SieveQ1 test data set. M5 model tree is employed to predict 1 hour ahead runoff. M5 model tree is also used to estimate uncertainty percentiles (prediction limits).

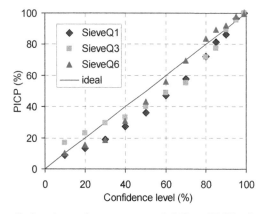

Figure 7.10. The prediction interval coverage probability (PICP) for different values of confidence level. M5 model tree is employed as the forecasting model and the uncertainty model in the UNEEC method.

Figure 7.9 shows the computed prediction limits for the 95% confidence level in the SieveQ1 test data set using the MT as primary (i.e. forecasting of Q_{t+1}) model and the uncertainty model in the UNEEC method. It can be seen that 96.67% of the observed data points are enclosed within the prediction limits. This value is very close to the desired value of 95%.

Figure 7.10 shows the deviation of the PICPs from the desired confidence level when the MT is used. The PIs are constructed for various confidence levels ranging from 10% to 99%. It is to be noticed that the PICPs are very close to the desired

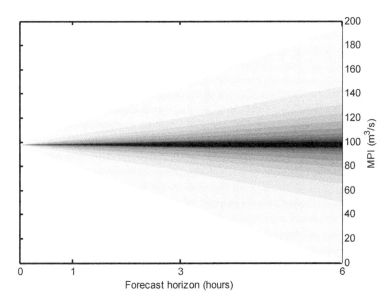

Figure 7.11. A fan chart showing the mean prediction interval for flow prediction in the Sieve catchment up to 6 hours ahead. The darkest strip covers 10% probability and the lightest - 99%.

confidence levels at values higher than 80%, and in practice the PIs are constructed around this value. Furthermore, it can be noted that the PIs are too narrow in most of the cases as the PICPs are below the straight line. Such evidence was also reported by Chatfield (2000). In these cases the UNEEC method underestimates the uncertainty of the model outputs.

Figure 7.11 presents a fan chart showing the MPI with different forecast lead times and different confidence levels. It is evident that the width of the PIs increases with the increase in the confidence level. Moreover it is also illustrated that the width of PIs increases as the forecast lead time increases, i.e. the model forecast uncertainty increases as the lead time increases.

7.2.5 Conclusions

This case study demonstrates the application of the UNEEC method for the prediction of uncertainty in the forecasts of river flows. We estimate uncertainty in the rive flow forecasts made by several data-driven techniques, namely artificial neural networks, model trees, and locally weighted regression. For this study, we select the Sieve River catchment located on the Appennines Mountains in Tuscany, North-Central Italy. We predict uncertainty for the forecast of 1, 3, and 6 hours lead time. The results shows that the UNEEC method gives reasonable estimate of the forecast uncertainty; and the percentage of observed data falling inside the uncertainty bounds estimated by the UNEEC method is very close to the desired degree of confidence level used to derive these bounds. We also show how uncertainty of forecast increases as forecast lead time increases.

7.3   Application 3: Brue catchment[4]

In this case study, we present the application of the UNEEC method to estimate the uncertainty of the simulated river flows of the Brue catchment in UK. A description of the Brue catchment is presented in section 5.1. The conceptual rainfall-runoff model HBV is used to simulate hourly river flows and its description is presented in section 5.2. The calibration of model parameters of the HBV model using adaptive clustering covering (ACCO) algorithm is described in section 5.3.

7.3.1   Analysis of simulation results

The analysis of the model residuals in the calibration period shows that model residuals are highly correlated with the observed flows. Most of the high flows have relatively high residuals whereas the low flows have small residuals. The presence of heteroscedasticity in the residuals is observed. Figure 7.12 shows the distribution of the residuals in the calibration and validation periods, and it can be seen that the residuals are not normally distributed. We have done several statistical tests such as the Kolmogorov-Smirnov and Lilliefors (Lilliefors, 1967) tests and these tests support this hypothesis. These tests of normality and homoscedasticity suggest that in order to provide a reliable estimate of the model uncertainty, transformations of the model residuals will be required if statistical methods are to be applied.

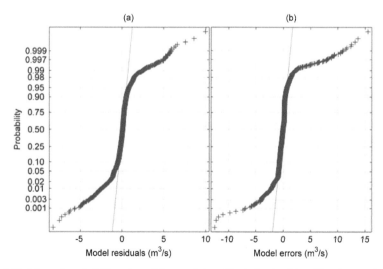

Figure 7.12. Normal probability plot of the model residuals in the (a) calibration period and (b) the model errors in the verification period.

---

[4] Based on: Shrestha, D.L. and Solomatine, D. (2008). Data-driven approaches for estimating uncertainty in rainfall-runoff modelling. *Intl. J. River Basin Management*, 6(2), pp. 109-122.

7.3.2 Selection of input variables

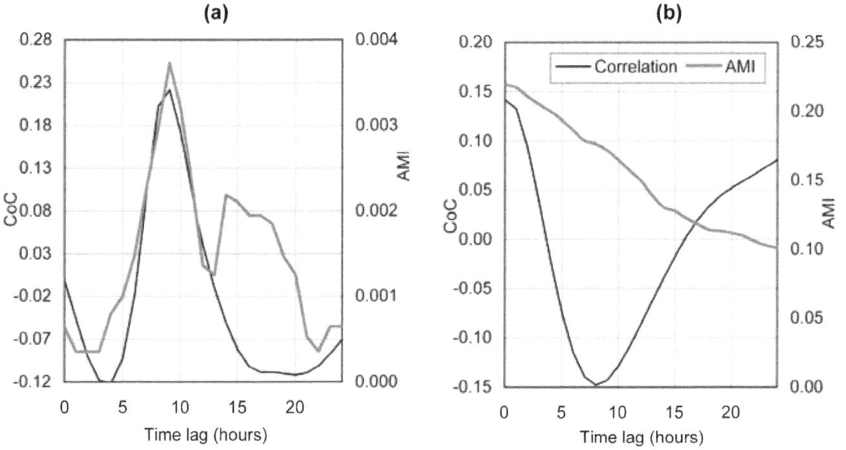

Figure 7.13. Average mutual information (AMI) and correlation of model residuals with (a) effective rainfall and (b) discharge.

Table 7.7. Summary of input variables selected for the models/processes.

Variables	HBV model	Clustering	Uncertainty model U
Input	R_t, E_t	RE_{t-8}, RE_{t-9}, RE_{t-10}, Q_{t-1}, Q_{t-2}, Q_{t-3}	RE_{t-8}, RE_{t-9}, RE_{t-10}, Q_{t-1}, Q_{t-2}, Q_{t-3}
Output	Q_t	V, P	PI_t^L, PI_t^U

Not: P is partition matrix that contains the membership values of each data for each cluster and V is the matrix of the cluster centres.

Several approaches have been reported in the literature (e.g., Guyon and Elisseeff, 2003; Bowden et al., 2005) to select the model input variables. We follow a similar approach which is also discussed in section 4.6. The input variables X_c used in the clustering consist of the previous values of effective rainfall (RE_{t-8}, RE_{t-9}, RE_{t-10}) and runoff (Q_{t-1}, Q_{t-2}, Q_{t-3}) (see Table 7.7). This structure of the input data used in the clustering is determined by the analysis of the correlation and average mutual information (AMI) between the effective rainfall or runoff and the model residuals (see Figure 7.13). One can notice that the model residuals have the highest correlation and AMI to the effective rainfall with the 9*th* time step lag. Whereas the residuals have the highest correlation and AMI to the discharge at the lag of zero. This is obvious because model residuals are computed from the observed and simulated discharges.

7.3.3 Clustering

Figure 7.14 shows the clustering of the input examples with the effective rainfall at a lag time of 9 hours (Figure 7.14a) and a discharge at a one hour lag (Figure 7.14b) in the abscissa and the model residuals in the ordinate. In this case study, we also set fuzzy exponential exponent m to 2. One can observe that high flows have a higher value of the model residuals, which are well identified by the clustering analysis. Furthermore, the

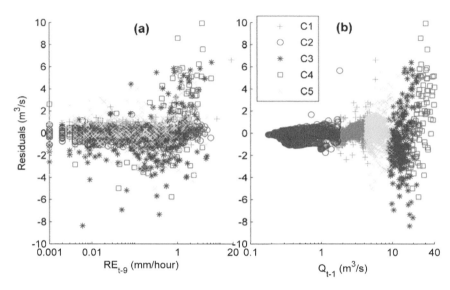

Figure 7.14. Plot of clustering results with the model residuals versus (a) effective rainfall and (b) discharge. The label C1 through C5 indicates the cluster ID.

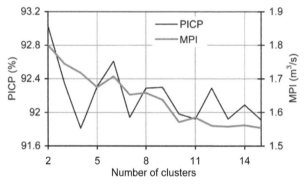

Figure 7.15. Sensitivity of the statistics of uncertainty measures with number of clusters.

low flows are also separated into one cluster, which has a very low value of the model residual.

An important issue is in determining the number of clusters. In this study, the number of clusters is determined by estimating the PIs with different numbers of clusters. The results are presented in Figure 7.15. We use the two statistics PICP and MPI (see section 6.6) in addition to visualize graphically the hydrograph in order to compare the results of the uncertainty analysis methods. In principle, these measures can also be computed individually for each cluster, but this would require training a classifier that would be able to attribute the new input vectors to one of the clusters.

Table 7.8. Cluster centres (only 2 variables are reported in this table) and computed prediction intervals for each cluster.

Cluster ID	Cluster centres		Prediction intervals	
	RE_{t-9} (mm/hour)	Q_{t-1} (m^3/s)	Lower	Upper
C1	0.059	2.727	-0.525	0.812
C2	0.026	0.518	-0.706	0.418
C3	0.433	12.712	-3.991	3.578
C4	1.451	26.311	-3.827	5.777
C5	0.14	5.849	-1.667	1.620

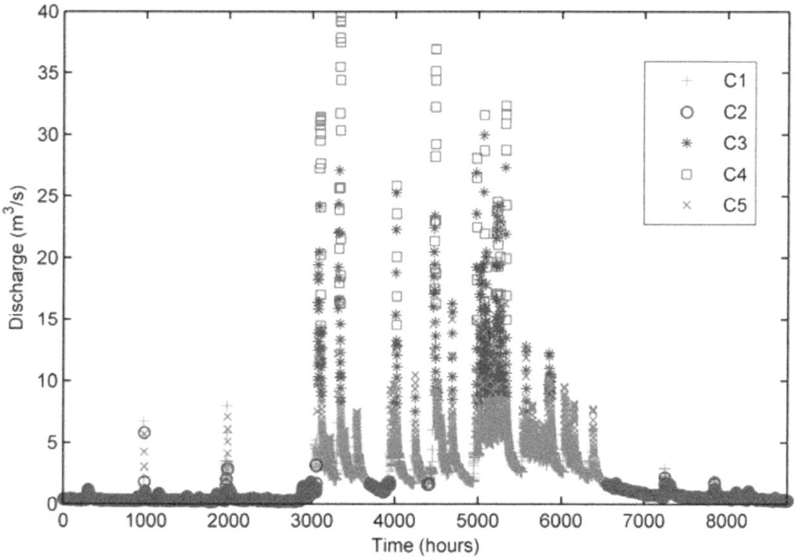

Figure 7.16. Fuzzy clustering of the input data for the calibration period (1994/06/24 05:00 - 1995/06/24 04:00).

It is observed that the number of clusters corresponding to PICP close to 90% confidence level is 4. However, MPI in this case is higher than that obtained with 5 clusters. So we choose to use 5 clusters. This value of 5 is also consistent with previous case study (section 7.2), which has shown that this value is reasonable to represent different situations related to the runoff generation process. The oscillation of PICP may be due to the following. The input data might be noisy. Empirical distributions estimated from the limited data may not be representative. The uncertainty model used in the prediction of the PIs may not be accurate.

Table 7.8 shows the computed PIs for each of the clusters with a 90% degree confidence level. Investigating the clusters with their centres and PIs reveals that the clusters with high values of rainfall and runoff have wider PIs, while the clusters with low values of rainfall and runoff have narrower PIs. The unsymmetrical values of the lower and upper PIs are obvious due to the fact that the HBV hydrological model calibration result was also biased. It is observed that most of the time the hydrological model is underestimating the river flows in the calibration period.

The fuzzy clustering of the input data for the calibration period is shown in Figure 7.16. The input data (in this figure the discharge data) is attributed to one of those clusters (Clusters C1 through C5) to which the data has the highest degree of membership. Cluster C2 contains input examples with very low runoff, whereas Cluster C4 is associated with very high values of runoff. Fuzzy C-means clustering is able to identify clusters corresponding to various mechanisms of the runoff generation process such as high flow, low flow, medium flow etc.

7.3.4 Uncertainty results and discussions

Figure 7.17 shows the estimated 90% PIs for the validation data using both regression and instance based learning methods as the uncertainty model. M5 model trees (Quinlan, 1992) is used as a regression model. In instance-based learning, we use a fuzzy committee approach and the weight is computed by setting $m = 1$ i.e., using inverse the square Euclidean distance (equation (6.7)).

It is found that when the instance-based method is used, 89.6% of the observed data points are enclosed within the computed PIs. In the case of the regression method a slightly higher share (90.8%) of the observed data is inside the PIs. 7.6% (6.8% with the regression method) of the validation data points fall below the lower PI, whereas only 2.8% (2.4% with the regression method) of the data points fall above the upper PI. This difference is consistent with the fact that the simulation model M is biased and overestimates the flow in the validation data. In order to check if the percentages of the bracketed data points are more or less similar for any range of river flows, we compare the histogram of the observed river flows, which are outside the bounds, with the observed ones. The result, which is not presented here, reveals that the distribution of the observed flows that are outside the uncertainty bounds is relatively consistent with the observed ones.

The average values of the PIs are 1.44 m³/s and 1.50 m³/s using the instance based and the regression methods respectively. Also, it can be noticed from Figure 7.17 that the PIs computed with the regression method are slightly wider than those obtained with the instance based method. These values are reasonable if compared with the order of magnitude of the model error in the test data (root mean squared error is 0.97 m³/s).

The upper part of Figure 7.17 depicts the width of the prediction bound and the fuzzy clustering of the data. Since in fuzzy clustering each data point belongs to all clusters (in this case clusters C1 through C5), the clusters to which each data point has the highest fuzzy membership values are identified and plotted. One can notice in the figure that the high flows have the highest fuzzy membership to the Cluster C5 and also have a large uncertainty, whereas the low flows have the highest fuzzy membership to the Cluster C2. The width of the prediction bound in this group is relatively small.

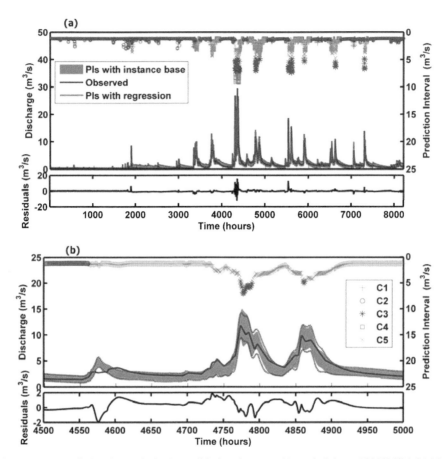

Figure 7.17. Prediction intervals in the validation data set. (a) Period from 1995/06/24 05:00 to 1996/05/31/ 13:00. Model residuals are shown in the bottom of the figure. The cluster ID and width of the PIs are also shown in the top axis of the figure. (b) Enlarge view for period from 1995/12/28 16:00 to 1996/01/18 08:00.

7.3.5 Comparison of uncertainty results

In this section the uncertainty results are compared with the widely used MC method GLUE (Beven and Binley, 1992) , and the meta-Gaussian method (Montanari and Brath, 2004). Note that the comparison with GLUE is performed purely for illustration since it analyses the parametric uncertainty. As far as the meta-Gaussian method is concerned, one thing is common to the UNEEC method - the uncertainty is based on the analysis of the optimal (calibrated) model residuals. The experiment for the GLUE method is setup as follows:

1. Prior feasible ranges of parameter values are set to be the same as those used in the automatic calibration of the HBV model (see Table 5.2);

2. The likelihood measure is based on the coefficient of the model efficiency, CoE criterion used also by Beven and Freer (2001);

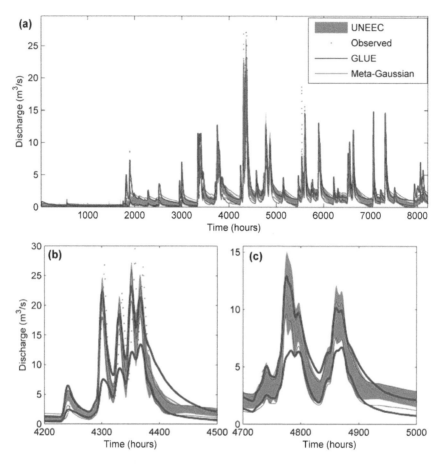

Figure 7.18. A comparison of 90% prediction bounds estimated with the UNEEC, GLUE, and meta-Gaussian approach in the validation period. (a) 1995/06/24 05:00 to 1996/05/31/ 13:00, (b) enlarge view of the period of about 12 days including the highest peak flow at 1995/12/22 and (c) the relatively small event after about one week later.

3. The rejection threshold values is set to 0.7; and

4. The number of behavioral parameter sets is set to 3000.

The comparison results are reported in Figure 7.18. One may notice the differences between the prediction bounds estimated with the UNEEC and GLUE method; nevertheless these two techniques are different in nature. The key difference is that GLUE as performed here accounts only for the parameter uncertainty whereas the UNEEC method treats all other sources of uncertainty in an aggregated form. Note that the width of the prediction bound obtained by the GLUE method varies with the rejection threshold and the likelihood measure to a great extent. For example, the lower rejection threshold produces relatively wider uncertainty bounds. The selection of appropriate rejection threshold is very crucial to the GLUE method, and it is one of the criticisms often reported in the literature.

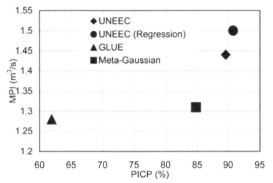

Figure 7.19. A comparison of the statistics of uncertainty estimated with the UNEEC, GLUE, and meta-Gaussian methods in the validation period. The statistics of the uncertainty are measured with PICP and MPI.

It is noticed that only 62% of the observed discharge in the validation data fall inside the 90% prediction bounds estimated by the GLUE method. As expected, the width of the prediction bounds is smaller than that obtained with the UNEEC method. The average value of the width of the prediction bound (i.e. MPI) is 1.28 m³/s. It appears that about an equal share of the observed data which are outside the prediction bounds are below (18.9%) and above (19.1%) the prediction bounds.

Figure 7.18 shows the 90% prediction bounds obtained with the meta-Gaussian approach in the validation data set. It is worthwhile repeating that the meta-Gaussian approach estimates the uncertainty based on the analysis of the optimal (calibrated) model residuals (see section 2.8.5). As we observe above that the model residuals are neither Gaussian nor homoscedastic, we transform the model residuals according to the outline presented by Montanari and Brath (2004) to stabilize the variance of the model residuals. It is observed that 84.8% of the observed discharge in the validation data falls inside the estimated 90% prediction bounds. Further analysis of the results reveals that 9% of the observed data are below the lower prediction bound whereas 6.2% of the data are above the upper prediction bound. The average width of the prediction bounds is 1.31 m³/s.

Figure 7.19 depicts the comparison statistics of the uncertainty estimation by the different methods considered. One can see that in the GLUE method most of observed data are outside the estimated prediction bounds with narrow uncertainty bounds compared to those obtained with the other methods. The 90% prediction bounds estimated with the UNEEC method encloses almost 90% of the observed data while a slightly smaller number of observed data are enclosed with the meta-Gaussian approach.

7.3.6 Conclusions

This case study presents the application of the UNEEC method for the prediction of uncertainty of the simulated river flows referring to the case study of the Brue catchment, UK. We use simplified version of the HBV conceptual model for the application of the method. Analysis of simulation results shows that the model residuals are neither normally distributed nor homoscedastic. Fuzzy clustering technique has been used to identify the natural groups of the input data. As a machine learning techniques model trees have been used to predict the uncertainty bounds of the model output simulated by the HBV model. Instance based learning method is also applied to predict the uncertainty bounds. The uncertainty results obtained from both methods are comparable.

The results show that the estimated uncertainty bounds are consistent with the order of magnitude of the model errors in the validation period. We also compare the uncertainty bounds with those estimated by the GLUE and the meta-Gaussian methods. Note that the width of the prediction bounds obtained by the GLUE method varies with the rejection threshold and the likelihood measure to a great extent. In the meta-Gaussian method, the model residuals should be normally distributed and homoscedastic to get reliable results. The comparison results shows that the UNEEC method gives reasonable estimate of the uncertainty of the model prediction; and the percentage of observed data falling inside the uncertainty bounds estimated by the UNEEC method is very close to the desired degree of confidence level used to derive these bounds. This application demonstrates that the UNEEC method can be used to predict the uncertainty of the model output.

7.4   Application 4: Bagmati catchment[5]

In this application, we demonstrate the application of the UNEEC method to a rainfall-runoff model of the Bagmati catchment in Nepal. The HBV model is also used as a rainfall-runoff model. Compared to the previous case studies, the Bagmati catchment has the following characteristics: the size of the catchment is bigger, the length of the data is larger, the resolution of the data is daily, and the quality of the data is comparatively poorer. In this study we also extend the methodology to estimate the probability distribution function of the model output.

7.4.1   Description of case study

Bagmati catchment lies in the central mountainous region of Nepal. The catchment encompasses nearly 3700 km² within Nepal and reaches the Ganges River in India. The catchment area draining to the gauging station at Pandheradobhan is about 2900 km² (see Figure 7.20) and it covers the Kathmandu valley, including the source of the Bagmati River at Shivapuri and the surrounding Mahabharat mountain ranges. The catchment covers eight districts of Nepal and is a perennial water body of Kathmandu. The length of the main channel is about 195 km within Nepal and 134 km above the gauging station.

Time series data for rainfall at three stations (Kathmandu, Hariharpurgadhi, and Daman) within the basin with a daily resolution for eight years (1988 to 1995) were collected. The mean areal rainfall is calculated using Thiessen polygons. Although this method is not recommended for mountainous regions, the mean rainfall is consistent with the long-term average annual rainfall computed with the isohyetal method (Chalise et al., 2003). The long-term mean annual rainfall of the catchment is about 1,500 mm with 90% of the rainfall occurring during the four months of the monsoon season (June to September). Daily flows are recorded from only one station at Pandheradobhan. Long-term mean annual discharge of the river at the station is 151 m³/s but the annual discharge varies from 96.8 m³/s in 1977 to 252.3 m³/s in 1987 (DHM, 1998). The daily potential evapotranspiration is computed using the modified Penman method recommended by the FAO (Allen et al., 1998).

Two thousand daily records from 1988/01/01 to 1993/06/22 are selected for calibration of the process model (in this study, the HBV hydrological model as described in section 5.2) and data from 1993/06/23 to 1995/12/31 are used for the validation (verification) of the process model. The first two months of the calibration data are considered as a warming-up period and are hence excluded from the study. This separation of the 8 years of data into calibration and validation data is done on the basis

[5] Based on: Solomatine, D. and Shrestha, D.L. (2009). A novel method to estimate total model uncertainty using machine learning techniques. Water Resources Research, 45, pp. W00B11.

Figure 7.20. Location map of Bagmati catchment considered in this study. Triangles denote the rainfall stations and circles denote the discharge gauging stations.

Table 7.9. Statistical properties of the runoff data sets

Statistical properties	Available data	Calibration set	Verification set
Period (yyyy/mm/dd)	1988/01/01 – 1995/12/31	1988/03/01 – 1993/06/22	1993/06/23 - 1995/12/31
Number of data	2922	1940	922
Average (m³/s)	152.73	140.15	179.17
Minimum (m³/s)	5.1	5.1	6.7
Maximum (m³/s)	5030.0	3040	5030
Standard deviation (m³/s)	273.31	226.42	350.83

of previous studies (Shrestha and Solomatine, 2005; Shrestha et al., 2006). These data sets are used with all the uncertainty analysis methods considered in this study. The some statistical properties of runoff data sets are presented in Table 7.9.

7.4.2 Calibration of model parameters

The HBV model is calibrated automatically using the global optimisation routine – adaptive cluster covering algorithm, ACCO (Solomatine et al., 1999) and followed by manual adjustments of the model parameters as done in the previous case study (see section 5.3). The HBV model parameters ranges used in the calibration procedure and their calibrated values are given in Table 7.10.

The Nash-Sutcliffe efficiency (CoE) value of 0.83 was obtained for the calibration period; this value corresponds to a root mean squared error (RMSE) value of 92.31 m³/s. The model was subsequently validated by simulating the flows for the independent validation data set. The CoE was 0.87 for this period with a root mean squared error value of 127.6 m³/s. Note that the standard deviation of the observed discharge in the validation period is 54% higher than that in the calibration period, and this apparently affects the increased performance in the validation period with respect to the CoE. The simulated and observed hydrographs, along with rainfall and simulation error, are shown in Figure 7.21.

Table 7.10. Ranges and calibrated values of the HBV model parameters. The uniform ranges of parameters are used for calibration of the HBV model using the ACCO algorithm and for analysis of the parameter uncertainty of the HBV model by the GLUE method.

Parameter	Description and unit	Ranges	Calibrated value
FC	Maximum soil moisture content (mm)	50-550	450
LP	Ration for potential evapotranspiration (-)	0.3-1	0.90
ALFA	Response box parameter (-)	0-4	0.1339
BETA	Exponential parameter in soil routine (-)	1-6	1.0604
K	Recession coefficient for upper tank (/day)	0.05-0.5	0.3
K4	Recession coefficient for lower tank (/day)	0.01-0.3	0.04664
PERC	Maximum flow from upper to lower tank (mm/day)	0-8	7.5
CFLUX	Maximum value of capillary flow (mm/day)	0-1	0.0004
MAXBAS	Transfer function parameter (day)	1-3	2.02

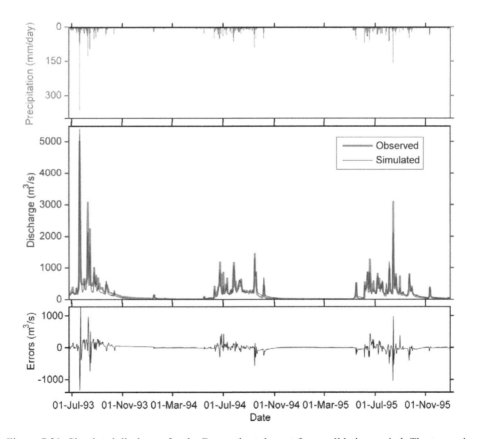

Figure 7.21. Simulated discharge for the Bagmati catchment for a validation period. The top and bottom plots show the precipitation and model errors respectively during the same period.

7.4.3 Analysis of model residuals

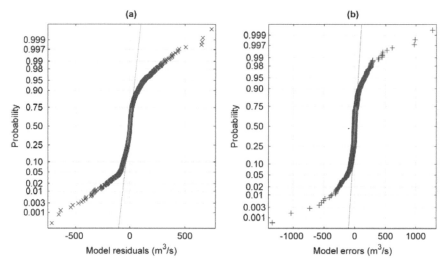

Figure 7.22. Normal probability plot of (a) model residuals in the calibration period and (b) model errors in the validation (or verification) period.

The analysis of the model residuals in the calibration period shows that the model residuals are highly correlated with the observed flows. Most of the high flows have relatively high residuals whereas the low flows have small residuals. The presence of heteroscedasticity in the residuals is observed as well. These results are consistent with those obtained in the Brue case study (section 7.3). Figure 7.22 shows the distribution of the residuals in the calibration and in the validation periods, and it can be seen that the residuals are not normally distributed. The Kolmogorov-Smirnov and Lilliefors (Lilliefors, 1967) tests support this hypothesis.

7.4.4 Clustering

The clustering is performed using Fuzzy C-means algorithm based on previous experience (Shrestha and Solomatine, 2006a, 2008). Selection of the input variables and the optimal number of clusters are discussed in this section.

Selection of input variables

Several approaches have been reported in the literature (Guyon and Elisseeff, 2003, Bowden et al., 2005) to select the model input variables; we follow the approach used in the previous case study (section 7.3). The input variables X_c used in clustering are constructed from the rainfall, the potential evapotranspiration, and the observed discharge. Several structures of the input data including the lagged variables are considered following the analysis of the correlation and the AMI between the rainfall, runoff and evapotranspiration with the model residuals. It appears that the inclusion of the potential evapotranspiration does not improve the results obtained for the cross-validation data set, and it can be said that its inclusion would introduce "confusion" into the model. For example, during the low flow season (i.e. the dry season of April and May) there is a very high potential evapotranspiration due to the hot weather in this period. The calibration of the hydrological model shows that the model captures the low flow reasonably well. However, this hydrological condition (low flow, negligible or zero rainfall, and very high potential evapotranspiration) is not identified as a low flow season in the clustering. So it was decided not to include the potential evapotranspiration as a separate variable, but rather to use the effective rainfall (see equation (5.10)).

After some trials with the different input combinations the following variables are selected for clustering: RE_t and Q_t (see Table 7.11). The principle of parsimony is followed by avoiding the use of a large number of inputs. In addition to this, the absolute values of the model residuals are used, which explicitly force the input data having similar values of model residuals to be in one group. Since the rainfall and discharge have different units with different orders of magnitude, and their values do not represent the same quantities, all input variables are normalized to the interval from zero to one. This scheme can prevent the model from being dominated by variables with large values, and is commonly used in data-driven modelling.

Table 7.11 Summary of input variables selected for the models/processes.

Variables	HBV model	Clustering	Uncertainty model
Input	R_t, E_t	RE_t, Q_t, ε_t	RE_t, RE_{t-1}, Q_t, Q_{t-1}
Output	Q_t	V, P	e_t^p

Note: P is the partition matrix that contains the membership values of data for each cluster and V is the matrix of the cluster centres.

Selection of number of clusters

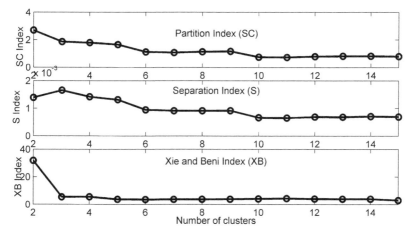

Figure 7.23. Finding the optimal number of clusters using partition index SC (top figure), separation index S (middle figure), and Xie-Beni index XB (bottom figure).

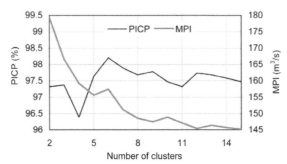

Figure 7.24. Sensitivity of the statistics of uncertainty measures with the number of clusters for the calibration period.

As previously mentioned, an important issue in clustering is to identify the optimal number of clusters. Three validation indices are used to select the optimal number of clusters: partition index (SC), separation index (S), and Xie-Beni index (XB) (Xie and Beni, 1991; Bensaid et al., 1996). The results are shown in Figure 7.23. The SC and S indices hardly decrease when the number of clusters c equals 6. However, the XB index reaches this local minimum at $c = 5$. It must be mentioned again, that the use of a single validation index is not reliable, so three indices are used, and the optimum can only be determined through a comparison of all the results. The partitions with fewer clusters are preferred when the differences between the values of a validation index are minor. So the optimal number of clusters is chosen to be 5, which is also confirmed by estimating prediction bounds for the calibration data (see Figure 7.24).

Figure 7.24 depicts the sensitivity of uncertainty measures to the number of clusters c. It is observed that MPI decreases with the increase of c. However, in the case of PICP there is no obvious pattern. The MPI fluctuates around the value 97.5% after $c=5$. At $c=5$, MPI and deviation of PICP from the desired confidence level (i.e., 90%) is smaller as compared to those with $c=6$. This value is also consistent with previous case studies, which have shown that this value is reasonable to represent different situations related to the runoff generation process.

Analysis of clusters

Figure 7.25 shows fuzzy clustering of input examples. The input variables, effective rainfall RE_t (Figure 7.25a) and discharge Q_t (Figure 7.25b), are on the abscissa and the model residuals are on the ordinate. Note that each input data belongs to all 5 clusters with different degrees of fuzziness (see Figure 7.26). However, in the plot the cluster which has the maximum membership function is shown. It is observed that there is a well defined pattern of the model residuals with the input variables such as defined by RE_t and Q_t. One can see that high flows and high (effective) rainfall generally have higher values of model residuals which are well identified (Cluster C3) by the clustering process. On the other hand, the conditions characterized by low flows are also separated into one cluster (Cluster C1) which has very low values of the model residuals. These results are also consistent with the results of the previous case studies.

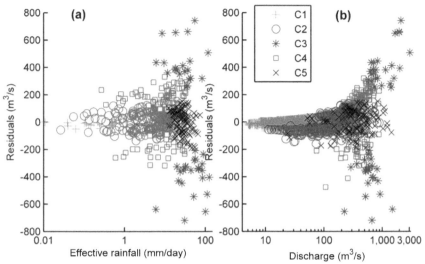

Figure 7.25. Fuzzy clustering of the input data in the calibration period (from 1988/01/01 to 1993/06/23) showing correlation of (a) effective rainfall and (b) discharge with model residuals. The labels C1 through C5 indicate the cluster ID.

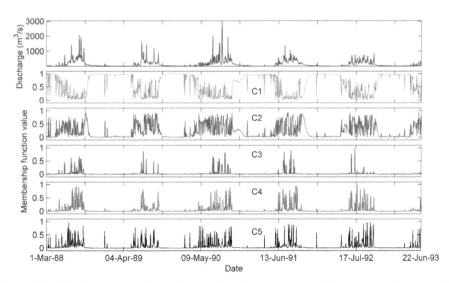

Figure 7.26. Membership functions of fuzzy clustering to the data. The top most plot shows the observed hydrograph. Each data point belongs to all Clusters C1, C2, C3, C4, and C5 with different membership function values.

Figure 7.27 presents the separation of the hydrograph with respect to the different clusters. One may notice that the majority of the flows have the highest membership in Cluster C1, which can be interpreted as the base flow. The peaks and most of the high flows are attributed to Cluster C3. Some high flows, especially on the recession of the hydrograph, are attributed to Cluster C5, because these examples have less or no rainfall

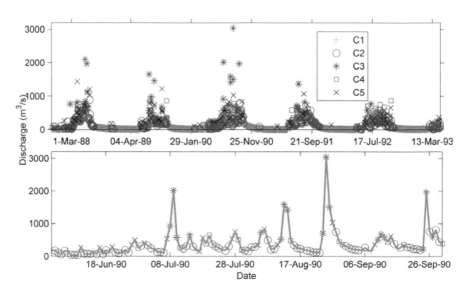

Figure 7.27. Fuzzy clustering of the input data in the calibration period (from 1988/01/01 to 1993/06/23). The bottom plot shows the enlarged view of the monsoon event of 1990.

and cannot be grouped into C3, which has high values of both flow and rainfall. It can be said that the Fuzzy C-means clustering is able to identify the clusters corresponding to the various mechanisms of the runoff generation process, such as peak flow with high rainfall, high flow with no or less rainfall (recession of the hydrograph) and base flow, etc.

7.4.5 Selection of input variables for uncertainty model

In order to select the most important influencing variables for the uncertainty model U, an approach similar to the one used for clustering is followed. Correlation analysis and AMI analysis between input variables RE_t and Q_t (including lags) and the quantiles of the model error are conducted Figure 7.28a shows the correlation coefficient and AMI of RE_t and its lagged variables up to 7 days, i.e., RE_{t-1}, RE_{t-2},..., RE_{t-7} with the 5% and 95% quantiles. It is observed that the variables RE_t and RE_{t-1} are strongly correlated with both quantiles; so these two variables are included in the input vector $\mathbf{x}_u$.

The correlation and AMI analyses between Q_t and the quantiles of the model error are presented in Figure 7.28b. It is also observed that Q_t and Q_{t-1} are strongly correlated with the quantiles. Although the lag 2 variable Q_{t-2} also has a high correlation, only Q_t and Q_{t-1} are included in the input vector. The reason is that the flow Q_t is highly auto-correlated and the inclusion of too many lagged variables of Q_t may lead to the redundancy of the model structure. Note that during the model application, Q_t is not available and we use its approximation made by model M. Although the simulated Q_t may bring additional uncertainty to the model U, our experiments have shown that this approach results in the more accurate model U (in terms of PICP and MPI).

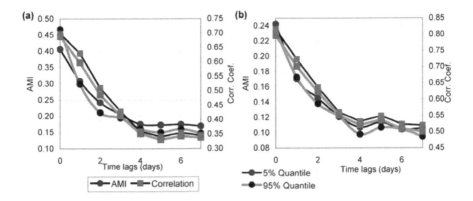

Figure 7.28. Average mutual information (AMI) and correlation coefficient (Corr. Coef.) of 5% and 95% quantiles of the model errors with (a) effective rainfall and (b) discharge. The thin dark line shows the 5% quantile, the thick grey line shows the 95% quantile.

Selection and validation of uncertainty model

M5 model tree (MT) is used as an uncertainty prediction model U. There are certain advantages of using MT compared to other machine learning methods; it is simple, easy, and fast to train. The results are interpretable, understandable, and reproducible. Solomatine and Dulal (2003) have shown that MT can be used as an alternative to ANN in rainfall-runoff modelling. There is only one parameter in MT, the pruning factor (or, alternatively, the minimum number of data allowed in each linear model component), which controls the complexity of the model. The following shows the structure of the input data for the model U to predict 5% and 95% quantiles (see also Table 7.11):

$$e^5 = U^5(RE_t, RE_{t-1}, Q_t, Q_{t\text{-}1};\ pf)$$

(7.13)

$$e^{95} = U^{95}(RE_t, RE_{t-1}, Q_t, Q_{t\text{-}1}; pf)$$

where e^5 and e^{95} are the 5% and 95% quantiles of the model error respectively, and pf is the pruning factor. The following is the example of a generated model tree by U^5 for e^5 with $pf = 4$:

$Q_t <= 73 : \text{LM1 (773)}$

$Q_t > 73 :$

$|\ \ Q_t <= 337 :$

$|\ \ |\ \ RE_t <= 8.73 :$

$|\ \ |\ \ |\ \ Q_{t-1} <= 255 : \text{LM2 (203)}$

$|\ \ |\ \ |\ \ Q_{t-1} > 255 : \text{LM3 (74)}$

$|\ \ |\ \ RE_t > 8.73 : \text{LM4 (77)}$

$|\ \ Q_t > 337 : \text{LM5 (173}$

LM1: $e^5 = 11.7 + 1.73RE_t + 0.909RE_{t-1} + 0.293Q_{t-1} + 0.302Q_t$

LM2: $e^5 = 44.7 + 0.282RE_t + 0.15RE_{t-1} + 0.0114Q_{t-1} + 0.198Q_t$

LM3: $e^5 = 80.4 + 0.282RE_t + 1.6RE_{t-1} + 0.0114Q_{t-1} + 0.0405Q_t$

LM4: $e^5 = 44.3 + 3.22RE_t + 0.167RE_{t-1} + 0.153Q_{t-1} + 0.0242Q_t$

LM5: $e^5 = 146 + 1.7RE_t + 0.0706Q_{t-1} + 0.00895Q_t$

There are five linear models (namely LM1, LM2, LM3, LM4, and LM5) generated for various intervals of Q_t, RE_t, and Q_{t-1}. Note that numbers inside the parenthesis are the numbers of the data vectors in the sub-sets. Similar structures of the linear models are obtained for e^{95}. Table 7.12 shows the mean and standard deviation statistics of the generated quantiles of the model error and performance of U^5 and U^{95} in the training, cross-validation and calibration data sets.

The mean and the standard deviation of the quantiles on the training and cross-validation data sets are consistent with the high variability in the original data. The performances of the uncertainty models U as measured by RMSE and CoE for 5% and 95% quantiles are quite reasonable for both the training and cross-validation data sets, in spite of the high variability of the quantiles. The RMSE and CoE values of the models U on the calibration data set are consistent with those for the training and cross-validation sets, and this ensures the predictability of the models U. Figure 7.29 depicts the scatter plot for the generated and predicted quantiles in the cross-validation data. It is observed that both models are quite good for approximating the relationship between the input space variables and the quantiles of the model error.

Table 7.12. Mean and standard deviation statistics of the 5% and 95% quantiles of the model error, and the performance of uncertainty prediction models U.

Data set	Mean	Std. dev.	RMSE	CoE
Training	73.60 (86.29)	77.54 (73.60)	28.44 (27.25)	0.87 (0.86)
Cross-validation	67.36 (79.58)	71.75 (67.70)	26.69 (27.41)	0.86 (0.84)
Calibration	71.55 (84.07)	75.72 (71.76)	27.74 (26.70)	0.87(0.86)

Note: Performances are RMSE and CoE. The training and cross-validation data constitutes 67% and 33%, respectively, of the calibration data (data used to calibrate process model). Values in parentheses correspond to statistics of 95% quantiles of the model error and the performance of uncertainty prediction models U^{95}. Std. dev., RMSE, and CoE are standard deviation of model errors, root mean squared error, the Nash-Sutcliffe between predicted and target values of the quantiles, respectively.

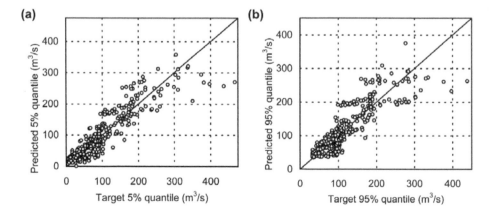

Figure 7.29. Scatter plot of the predicted and the target quantiles of the model errors for (a) 5% and (b) 95% quantiles in cross-validation data (part of the calibration data).

Analysis of model uncertainty

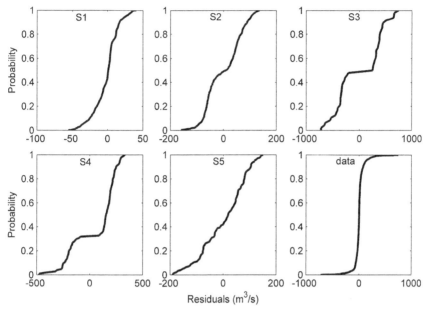

Figure 7.30. Cumulative distribution of the model residuals for each cluster in the calibration period. The last figure (second row and third column in the figure panel) shows the cumulative distribution of the model residuals for the calibration data.

Table 7.13. Cluster centres and computed quantiles of the model error for each cluster.

Cluster	Cluster centre		Quantiles	
	RE_t (mm/day)	Q_t (m³/s)	5%	95%
C1	0.39	33.99	-43.27	32.28
C2	2.89	176.88	-91.44	107.71
C3	39.61	792.44	-530.47	442.49
C4	10.21	469.22	-264.14	288.29
C5	27.02	385.45	-184.77	189.90

Note: Cluster C1 is identified by low values of flows and rainfall. Cluster C3 is identified by high flows and high rainfalls.

Figure 7.30 shows the empirical cumulative distribution of the model residuals for each cluster. In this plot sets S1, S2, S3, S4 and S5 belong to the sets of the input data which has the highest membership grade to the Cluster C1, C2, C3, C4 and C5, respectively. Obviously the distributions are quite different among them.

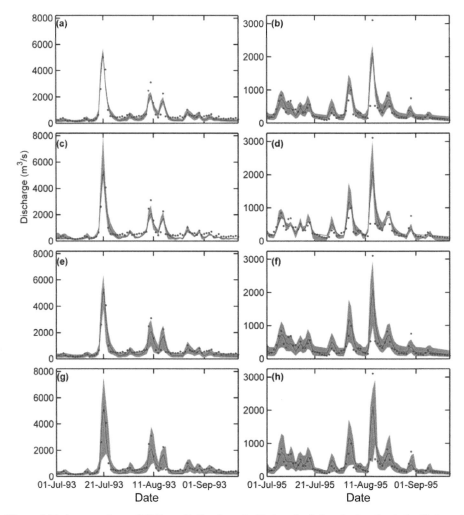

Figure 7.31 A comparison of 90% prediction bounds (darker shaded region) estimated with (a and b) the UNEEC method, (c and d) the GLUE method, (e and f) the meta-Gaussian method and (g and h) the quantile regression method in the validation period. The figures on the left shows the monsoon period of 1993 and the figures on the right show the monsoon of 1995. The dots show the observed discharge, the line shows the simulated discharge.

5% and 95% quantiles of the model residuals are computed for each of the clusters. The results are presented in Table 7.13. Investigating the clusters with their centres and quantiles reveals that the clusters with high values of rainfall and runoff have large values of the quantiles, while the clusters with low values of rainfall and runoff have small values of the quantiles. The asymmetrical values of these quantiles are obvious due to the fact that the HBV hydrological model calibration result is biased. It is observed that most of the time the hydrological model underestimates the river flows in the calibration period.

Figure 7.31 shows the observed discharge, the 90% hydrograph prediction uncertainty and comparison to the other three methods. The details of the comparison follow later. Figure 7.31a highlights the flood event that occurred during the monsoon of 1993. Interestingly enough the HBV model captures the highest peak flow very well and consequently the peak flow is bracketed by the predicted PIs. Figure 7.31b focuses on another monsoon event during 1995. This event was underestimated by the HBV model. One can see that the estimated uncertainty bound fails to enclose the highest peak discharge of the 1995 monsoon.

It is observed that 88.07% of the observed data points are enclosed within the computed PIs. 6.4% of the validation data points fall below the lower PI, whereas 5.53% data points fall above the upper PI. The average width of the uncertainty bounds i.e. MPI is 165 m^3/s. This value is reasonable if compared with the order of magnitude of the model error in the validation data. The further analysis reveals that the distribution of the observed discharge below the lower PI is relatively consistent with the observed discharge. As far as the upper PI is concerned, less data are outside in the low flow (range of 0-250 m^3/s). This means that the upper PIs are unnecessarily overestimated. However, the width of the upper PI in the intermediate flows (range of 250-750 m^3/s) is considerably narrower.

7.4.6 Comparison of uncertainty results

In this section the results are compared with the widely-used Monte Carlo method GLUE, the meta-Gaussian, and the quantile regression (Koenker and Bassett, 1978) method. Note that the comparison with GLUE is performed purely for illustration since it analyses the parametric uncertainty, and other methods – the uncertainty based on the analysis of the optimal (calibrated) model residuals.

The experiment for the GLUE method (Beven and Binley, 1992) is set up similar to the Brue case study except, the number of behavioural parameter sets is increased to 25,000. The comparison results are reported in Figure 7.31c, d. One may notice the differences between the prediction bounds estimated by the UNEEC and GLUE method, but, again, these two techniques are different in nature, which was discussed in the previous case study (section 7.3).

It can be noticed that only 63.9% of the observed discharge values in the validation data fall inside the 90% prediction bounds estimated by the GLUE method. As expected, the width of the prediction bounds is smaller than those obtained with the other methods. The average value of the prediction bound width is 120.35 m^3/s (see Figure 7.32). The further detailed analysis reveals that only 6.51% of the observed discharges are below lower PIs. The majority of the observed flows (29.61%) fall above the upper PIs.

The 90% prediction bounds estimated with the meta-Gaussian approach are shown in Figure 7.31e, f. Interestingly, it is found that 90% (more accurately, 90.02%) of the observed discharge values in validation data fall inside the estimated 90% prediction bounds (see also Figure 7.32). Further analysis of the results reveals that 3.7% of the

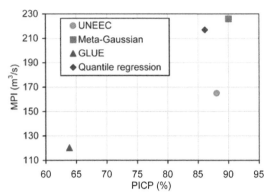

Figure 7.32. A Comparison of the statistics of uncertainty estimated with the UNEEC, meta-Gaussian, GLUE, and quantile regression methods in the validation period.

observed data are below the lower PI whereas 6.3% of data are above the upper PI. However, one can see that the bounds width is consistently larger and the average width of the prediction bounds is 225.79 m³/s, which is about 35% larger than that estimated by the UNEEC method.

Quantile regression method was used to compute the 5% and 95% error quantiles. The input variables **x** (see equation (2.26)) used in this method are the same as those used in the UNEEC method. We have also made several experiments with other input variables combination, but the results are not so good. The best results are reported in Figure 7.31g, h. It is observed that 86.12% of the observed discharge values in validation data fall inside the estimated 90% prediction bounds (see Figure 7.32). Further analysis of the results reveals that 4.01% of the observed data are below the lower PI, whereas 9.87% of data are above the upper PI. The average width of the prediction bounds is 216.86 m³/s.

7.4.7 Estimation of probability distribution of model errors

In the previous sections, only 90% prediction intervals, i.e., 5% and 95% quantiles are estimated. In this section the extension of the method to derive a more accurate estimate of the pdf is demonstrated. Several quantiles (such as 2.5, 5, 10:10:90, 95, 97.5%) are computed for the calibration data after clustering of the input space. Then regression models are trained for each quantile independently:

$$e^p = U^p(RE_t, RE_{t-1}, Q_t, Q_{t-1}; \; pf) \quad (7.14)$$

where p = 2.5, 5, 10:10:90, 95, 97.5 %. In this experiment, a total of 13 regression models are trained. Once the models U^p are trained on the training data, they are used to predict the quantiles for the new input data (e.g., for the validation data set). Note that since MTs are used as regression models, it takes only a couple of seconds to train a single model, so the computational cost to estimate the full distribution is not a major concern. However, this could be an issue for computationally extensive algorithms, such as support vector machine or ANNs with long records of input data.

The cumulative probability distribution (cdf) for the peak discharge of the flood dated 21 July 1993 (the highest peak event of the 1993 monsoon shown in Figure 7.31) is shown in Figure 7.33a. The cdf computed from the GLUE, meta-Gaussian, and quantile regression methods are also presented for comparison. One can notice that the cdf computed from the UNEEC method is relatively steep. The GLUE and quantile regression methods produce a comparatively flat cdf. For this particular flood event this means that the uncertainties estimated by the GLUE and quantile regression methods are very high and the uncertainty estimated by the UNEEC method is lower. The meta-Gaussian method gives intermediate results. Additional analysis of the cdf for the flood event of 14 August 1995 supports the finding that the uncertainty estimated with the UNEEC method is consistently lower for the flood events (Figure 7.33b).

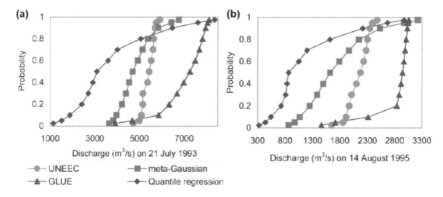

Figure 7.33. A Comparison of estimation of cumulative probability distribution for peak discharges of the monsoon period of (a) 1993 and (b) 1995.

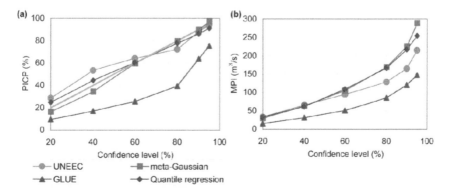

Figure 7.34. A Comparison of the statistics of uncertainty measures. (a) Prediction interval coverage probability (PICP). In an ideal case, the plot between PICP and confidence level follows the thick grey line. (b) Mean prediction interval (MPI) for different values of the confidence level.

Further analysis is performed in order to compute the percentage of the observed discharge data falling within the estimated uncertainty bounds (i.e. PICP) and the average width of the uncertainty bounds (i.e. MPI) for various specified confidence levels (ranging from 20% to 95%) used to produce these uncertainty bounds. The results are presented in Figure 7.34. As far as the PICP is concerned, the ideal would be to follow the thick grey line (Figure 7.34a). Points below this line indicate that less data are bracketed by the uncertainty bounds. On the other hand, if more data are enclosed in the uncertainty bounds, the PICP line would be above the ideal line. It can be seen that the PICPs computed with the meta-Gaussian and QR methods are very close to the desired confidence levels. In the UNEEC method more data are enclosed at lower values of the confidence levels. The GLUE method produces consistently lower values of PICPs for all confidence levels. The results with the GLUE method are consistent with the results reported in the literature, namely that the percentages of the observed discharge falling within the uncertainty bounds estimated by the GLUE is normally much lower than the specified confidence levels (see, e. g., Montanari, 2005; Xiong and O'Connor, 2008). The low values of PICP of the GLUE uncertainty bounds can be attributed to many factors which are discussed in section 5.6.2.

As far as the MPI is concerned, the GLUE method generates relatively narrower uncertainty bounds (Figure 7.34b). The MPI values estimated by UNEEC and meta-Gaussian methods are comparable at the lower values of the confidence levels. However, uncertainty bounds estimated by the meta-Gaussian method increase faster as compared to those obtained with the UNEEC method after 60% confidence levels. The MPI estimated with the QR method is similar to that obtained with the meta-Gaussian method.

7.4.8 Conclusions

This case study presents the application of the UNEEC method for the prediction of uncertainty of the simulated river flows referring to the case study of the Bagmati catchment in Nepal. In this case study, we also observe that the model residuals are neither normally distributed nor homoscedastic. Compared to the case study of the Brue catchment, the uncertainty of the model prediction is higher; this is obvious due to the following reasons: (i) the size of catchment of this case study is larger and thus basin average rainfall, temperature data may not be representative; (ii) the quality of data is poorer; and (iii) the resolution of data is higher.

The results show that the estimated uncertainty bounds are consistent with the order of magnitude of the model errors in validation period. We also compare the uncertainty bounds with those estimated by GLUE and meta-Gaussian and quantile regression methods. The comparison results shows that the UNEEC method gives reasonable estimate of the uncertainty in the model prediction; and the percentage of observed data falling inside the uncertainty bounds estimated by the UNEEC method is very close to the desired degree of confidence level used to derive these bounds. It has been also demonstrated the extension of the UNEEC method to approximate the probability distribution function of the model output.

7.5   Multiobjective calibration and uncertainty

In sections 7.1-7.4, uncertainty analysis of the optimal model is carried out. The model is calibrated using the single objective function value. This section presents some results with multiobjective calibration and uncertainty.

Practical experience with the calibration of the hydrological model suggests that a single objective function value is often inadequate to measure properly the simulation of all the important characteristics of the system that are reflected in the observations. Gupta et al. (1998) pointed out that there may not exist an objective "statistically correct" choice for the objective function, and therefore no statistically correct "optimal choice for the model parameters. Furthermore, recent advances in computational power have led to more complex hydrological models, often predicting multiple hydrological fluxes simultaneously. These issues have led to an increasing interest in the multi - objective calibration of hydrological model parameters (Gupta et al., 1998; Yapo et al., 1998; Madsen, 2000; Khu and Madsen, 2005).

Multiobjective calibration problem can be stated as follows:

$$\min\{f_1(\theta),...,f_m(\theta)\}$$

$$(7.15)$$

where m is the number of objective functions, $f_i(\theta)$, $i = 1, ..., m$ are the individual objective functions, and θ is the set of model parameters to be calibrated. Due to trade-offs between the different objectives, the solution to equation (7.15) will no longer, in general, be a single unique parameter set. Instead, there exist several solutions that constitute a so-called Pareto optimal set or non-dominated solutions. Any solution θ^* belongs to the Pareto set when there is no feasible solution θ that will improve some objective values without degrading performance in at least one other objective. Mathematically, the solution θ^* is Pareto-optimal (i) if and only if $f_k(\theta^*) \leq f_k(\theta)$ for all $k = 1, ..., m$ and (ii) $f_j(\theta^*) < f_j(\theta)$ for some $j = 1, ..., m$. According to these two statements the Pareto-optimal solution θ^* has at least one smaller objective value compared to any other feasible solution θ in the decision space, while performing as well or worse than θ in all remaining objectives.

The multiobjective optimisation algorithm used in this study is the non-dominated sorting genetic algorithm (NSGA-II) developed by Deb et al. (2002). NSGA-II is capable of handling a large number of objective functions and provides an approximate representation of the Pareto set with a single optimisation run. The NSGA-II algorithm is outlined briefly as follows:

1.   Generate an initial population of size p randomly in the domain of the feasible range;

2.   Evaluate the population and sort the population based on non-domination using a bookkeeping procedure to reduce the order of the computation;

3.   Classify the population into several Pareto fronts based on the non-domination level. Individuals belonging to the first Pareto front are assigned with rank 1,

individuals belonging to the second Pareto front (second Pareto front is the Pareto front after removing the individuals from the first front) are assigned with rank 2, and so on;

4. Form a new population by generating an offspring population from the parents and combining them with the parents (following standard GA procedure);

5. Compute the crowding distance for each individual;

6. Select the individuals based on a non-domination rank. The crowding distance comparison is made if the individual belongs to the same rank;

7. Repeat the steps 3 – 6 until the stopping criterion is satisfied. The stopping criterion may be a specified number of generations, maximum number of function evaluations or computation time.

7.5.1 Preference ordering of the Pareto-optimal points

It is observed that the number of the Pareto-optimal solution grows with the dimensionality of the objective functions used in the multiobjective optimisation. It was reported that the qualitatively of the solutions are indistinguishable from each other. However, it is often required to select a Pareto-optimal point from a set of many and it becomes highly intractable when the number of solutions becomes large. For example, the number of Pareto-optimal solutions in calibrating hydrological model parameters in four dimension objective functions reaches easily above 100. Since the decision maker generally prefers a small numbers of solutions, modelers have to face the task of selecting a set of suitable model parameters from the numerous Pareto-optimal sets.

The concept of the preference order of efficiency, which was first introduced by Das (1999), can be applied to select the most preferable solutions by setting up a hierarchical ordering among the various solutions obtained from the multiobjective optimizations. The preference ordering (PO) uses the following two theorems (Das, 1999), which are strong domination criteria compared to Pareto domination:

The first theorem, efficiency of order k (k-Pareto-optimal points), considers all the possible k-dimensional subspace of the original m-dimensional objective function space ($1 \leq k \leq m$). A solution is defined as being efficient of order k, if this solution is not dominated by any other solution in any of the k-dimensional subspaces. The second theorem is the efficiency of order k with degree p (denoted as $[k, p]$). A solution is defined as being efficient of order k with degree p, if it is not dominated by any other solutions for exactly p out of the possible $_mC_k$ k-dimensional subspaces.

The condition of efficiency of order can thus be used to help reduce the number of solutions from the Pareto-optimal set by retaining only those that are regarded as the best compromises. When the number of points selected is still considerably large, a more stringent condition of efficiency with degrees is required to sort out better solution. Thus by reducing the efficiency of order and increasing the degree of order in a sequential manner, the criteria of Pareto-optimality is tightened, which enables the modeller to sieve through the large number of Pareto-optimal sets and determine the preferred solution (Khu and Madsen, 2005).

7.5.2 Results and discussions

The NSGA-II algorithm was applied to Bagmati catchment for multiobjective calibration of HBV model parameters. Four different model performance measures that emphasise different aspects of the hydrograph are considered as objective functions for calibration. These objective functions are as follows:

$$\text{Volume error: } f_1(\theta) = \left| \sum_{i=1}^{N} [Q_i - \hat{Q}_i] \right| \tag{7.16}$$

$$\text{Root mean square error (RMSE): } f_2(\theta) = \left[\frac{1}{N} \sum_{i=1}^{N} [Q_i - \hat{Q}_i]^2 \right]^{1/2} \tag{7.17}$$

$$\text{Low flow RMSE: } f_3(\theta) = \left[\frac{1}{N_l} \sum_{i=1}^{N_l} [Q_i - \hat{Q}_i]^2 \right]^{1/2} \tag{7.18}$$

$$\text{High flow RMSE: } f_4(\theta) = \left[\frac{1}{N_h} \sum_{i=1}^{N_h} [Q_i - \hat{Q}_i]^2 \right]^{1/2} \tag{7.19}$$

where Q_i is the observed discharge at time i, $\hat{Q}_i$ is the simulated discharge, N is the number of time steps in the calibration period, N_l is the number of time steps in low flow events, N_h is the number of time steps in high flow events. In this study, low flow events are defined as periods with flow below a threshold value of 50 m³/s and high flow events are defined as periods with flow above 500 m³/s.

Since GA is a stochastic process, 10 independent calibration runs are carried out to compare the quality of the solutions of each run. It is observed that each independent run is comparable to the other. The maximum number of the generation and the population size is set at 50 and 80, respectively. The probability of crossover and mutation are set to 0.9 and 0.1, respectively. We also use a single objective GA algorithm to calibrate each of the single objective functions, (equations (7.16) - (7.19)), i.e., optimising for volume error $f_1(\theta)$, RMSE $f_2(\theta)$, low flow RMSE $f_3(\theta)$, high flow RMSE $f_4(\theta)$. The following results are from one of the ten runs.

Table 7.14 presents the number of points on the Pareto front for all possible combinations of the four objective functions (equations (7.16)-(7.19)). Since there are four objective functions ($m=4$), the possible combinations are given by $_mC_k$, where $K = $

Table 7.14. All possible combinations of the four performance measures, denoted as 1-2-3-4 and the resulting number of Pareto-optimal solutions.

Runs	1-2-3-4	1-2-3	1-2-4	1-3-4	2-3-4	1-2	1-3	1-4	2-3	2-4	3-4
1	413	349	33	362	88	17	67	18	33	11	39
2	347	260	33	289	73	13	46	8	30	12	41
3	392	315	41	318	70	17	50	12	31	12	27
4	300	240	28	255	62	13	48	8	27	7	35
5	425	327	46	328	81	17	44	12	37	8	41
6	342	296	32	275	56	13	39	9	29	4	24
7	371	303	47	321	64	12	52	11	34	14	36
8	387	340	27	331	47	11	46	10	32	4	34
9	294	231	41	229	79	14	55	16	36	12	41
10	382	335	38	339	43	14	54	14	28	10	37
Avg.	365	300	37	305	66	14	50	12	32	9	36

{4, 3, 2}. In total there are 11 combinations – (i) one combination of all four objective functions (denoted as (1-2-3-4); (ii) four combinations of any three objective functions (i.e., (1-2-3), (1-2-4), (1-3-4), (2-3-4)); (iii) six combinations of any two objective functions (i.e., (1-2), (1-3), (1-4), (2-3), (2-4), (3-4)). It can be seen from Table 7.14 that the numbers of solutions for each of the 10 runs are not significantly different from each other. Although this does not guarantee that the good solutions are present, it does indicate that the quality of each run is comparable to the others.

It is observed that the resulting number of Pareto-optimal points in 4-dimensional space is, on average, about 365. With such a large number of Pareto-optimal solutions, it is not feasible to examine each and every one of the solutions to identify and select the most appropriate calibration parameter set. One must therefore be able to sieve through these solutions to obtain a smaller number of representative solutions. If we examine the four possible combinations of any three objective functions (columns 3– 6), the resulting number of Pareto-optimal points ranges between 37 and 305. If we examine the six possible combinations of any two objective functions, the resulting number of Pareto-optimal points ranges between 9 and 50. However, it should be noted that these Pareto-optimal points for the two objective functions may not be Pareto-optimal in the other objective function spaces.

Preference ordering has been applied to select a small number of solutions from the higher dimension Pareto-optimal solutions. The results are presented in Table 7.15. As the efficiency of order (k) decreases, the number of preferred solutions decreases; and as the degree (p) of that order increases, the number of preferred solutions decreases. From Table 7.14 we see that the number of solutions reduces from an average of 361 to 2 when the degree of order 3 increases from 1 to 4. Thus the problem becomes highly manageable when the number of solutions becomes small while the qualities of these solutions in terms of Pareto-optimality are good (they are of a high degree and high order of efficiency, i.e., [3,4]).

Taking the example of run 1 in Table 7.15, it can be seen that if all four objective functions are considered, there are a total of 413 solutions. However, a closer examination of these solutions indicates that not all of them are Pareto-optimal if the

Table 7.15. Results of applying preference ordering to determine the number of preferred solutions in the 4th, 3rd, and 2nd order of efficiency (k) for different degrees (p) denoted by [k, p].

Runs	[4,1]	[3,1]	[3,2]	[3,3]	[3,4]	[2,1]	[2,2]	[2,3]	[2,4]	[2,5]	[2,6]
1	413	410	360	61	1	146	33	6	0	0	0
2	347	342	275	35	3	123	23	4	0	0	0
3	392	389	304	50	1	113	28	8	0	0	0
4	300	298	246	41	0	105	26	7	0	0	0
5	425	418	309	55	0	118	34	7	0	0	0
6	342	334	281	42	2	87	24	7	0	0	0
7	371	367	323	43	2	121	29	9	0	0	0
8	387	382	318	43	2	103	30	4	0	0	0
9	294	289	244	45	2	142	28	4	0	0	0
10	382	377	327	48	3	116	34	6	1	0	0
Avg.	365	361	299	46	2	117	29	6	0	0	0

objective function combinations are reduced and changed. When the combinations of any three out of the four objective functions are considered, (i) 410 of the original 413 Pareto-optimal solutions are present in at least one of the four possible combinations; (ii) 360 of the original 413 solutions are present in at least two of the four combinations; (iii) 61 of the 413 solutions are present in at least three of the four combinations; and (iv) one of the 413 solutions are present in all four combinations. Thus the preference ordering techniques has selected one preferred solution out of the original 413 Pareto-optimal solutions. If the number of solutions has not been reduced sufficiently, the order of efficiency should be reduced to 2 instead of 3 and the degree of efficiency be considered in turn. The remaining columns in Table 7.15 show such a process ([2,1], [2,2], [2,3], [2,4], [2,5], [2,6]). For run 1 no solutions are present in degrees higher than 3 (i.e., four or more of the six two-objective combinations). Thus, in this case the one [3,4] Pareto-optimal solutions are selected for further analysis.

Table 7.16 shows the comparison of the parameter values of the preferred [3,4] Pareto-optimal set and the optimal values of the parameter for each of the single objective functions. As expected, the values of each parameter vary depending on the selected objective functions used for calibration. Note that the values of the parameters of the preferred Pareto-optimal set are within the ranges of the parameter values of the single objective calibrations except first two parameters. The variation of the optimal model parameters sets of four single objective calibrations along the Pareto front is shown in Figure 7.35.

Table 7.17 presents the comparison of the objective function values of the preferred Pareto-optimal set and the objective function values for each of the single objective optimisation. The widely used Nash-Sutcliffe coefficient (CoE) is also calculated as an additional performance measure. As expected, the best calibration and validation results for each objective functions are obtained using single objective functions optimisation. For example, the lowest RMSE value of 7.00 m^3/s on low flow events is obtained when the parameters are optimised for single objective function $f_3(\theta)$. Similarly, the lowest RMSE value of 265.78 m^3/s on high flow events is obtained via calibrating objective function $f_4(\theta)$. The results of the Pareto-optimal set are a good compromise between all

Table 7.16. Comparison of the preferred model parameters with the optimal parameters corresponding to the four single objective function values.

Parameters	Preferred solution	Optimised for single objective function			
		f_1	f_2	f_3	f_4
FC	549.565	199.955	547.500	538.892	541.998
LP	0.987	0.831	0.810	0.301	0.830
ALFA	0.195	0.063	0.316	1.221	0.170
BETA	1.093	1.000	1.013	4.739	1.000
K	0.219	0.105	0.128	0.053	0.278
K4	0.068	0.300	0.049	0.019	0.275
PERC	6.104	0.364	7.019	3.038	6.977
CFLUX	0.226	0.057	0.315	0.086	0.184
MAXBAS	2.004	1.506	2.067	1.042	2.153

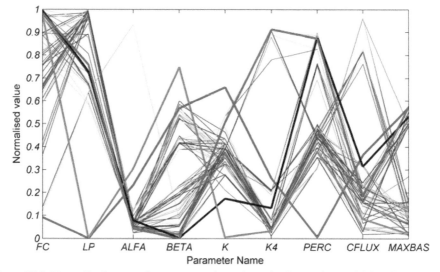

Figure 7.35. Normalised range of parameter values along the Pareto front. Thick solid magenta, black, green, and red lines indicate parameter sets calibrated (single objective) for volume error, RMSE, low flow RMSE, and high flow RMSE, respectively.

four objective functions and are within the best and worst solutions from the single objective function optimisations.

Figure 7.36 shows the Pareto sets (best 80 solutions) of various two-objective function combinations. Note that all these solutions are from Pareto sets of optimisation results using all four objective functions. For comparison, the solutions obtained from single objective optimisations and the preferred Pareto-optimal set are also shown in the figure. Further analysis of correlation between the objective functions shows that the objective function $f_2(\theta)$ (i.e. RMSE) is highly correlated (linear correlation coefficient of 0.99) with the objective function $f_4(\theta)$ (i.e. high flow RMSE). There is little positive correlation (correlation coefficient of about 0.38) between objective function $f_1(\theta)$ (volume error) and $f_2(\theta)$. As expected, the objective functions $f_1(\theta)$ and $f_4(\theta)$ are also

Table 7.17. Comparison of the ranges of model performances corresponding to the Pareto font with the four single objective function values in calibration and verification period

Performance measures	Preferred solution	Optimised for single objective function			
		f_1	f_2	f_3	f_4
Calibration Data					
Volume error %, f_1	1.80	**1.55E-5**	5.22	10.86	3.88
RMSE, f_2	93.14	123.48	**91.31**	223.74	99.34
Average low flow RMSE, f_3	27.24	35.83	22.65	**7.00**	35.31
Average peak flow RMSE, f_4	270.67	381.53	279.13	678.91	**265.78**
Nash-Sutcliffe CoE	0.83	0.70	**0.84**	0.02	0.81
Verification Data					
Volume error %, f_1	0.18	**0.0533**	1.85	7.36	3.90
RMSE, f_2	99.25	126.6861	**95.07**	276.48	102.21
Average low flow RMSE, f_3	24.47	26.287	**21.05**	36.00	31.25
Average peak flow RMSE, f_4	303.06	391.556	305.40	881.58	**284.86**
Nash-Sutcliffe CoE	0.80	0.6791	**0.82**	-0.53	0.79

Note: Bold type indicates the best value.

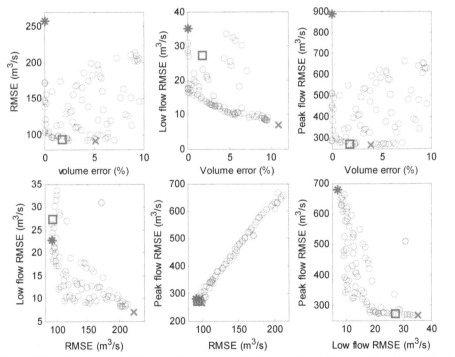

Figure 7.36. Various two-objective function combination plots of Pareto-optimal sets. Circle indicates Pareto solutions in 4-D objective functions space, plus marks in circles indicate Pareto solutions in 2-D space. Asterisk and cross indicates solutions corresponding to the single objective calibration and square indicates the preferred Pareto-optimal set.

correlated with an almost equal value of the correlation coefficient (0.39). However, as seen in the figure, other two-objective functions combinations such as $f_1(\theta)$ and $f_3(\theta)$ or $f_2(\theta)$ and $f_3(\theta)$ or $f_3(\theta)$ and $f_4(\theta)$ have negative correlation (correlation coefficient ranges from 0.53-0.60). The negative correlation means the two objective functions are conflicting.

The simulated hydrographs obtained by multi and single objective optimisations in both calibration and validation periods are shown in Figure 7.37. The dark shaded region is the range of the simulation obtained by the best 80 Pareto-optimal sets. It is observed that simulations obtained by single objective optimisations are within the range of hydrographs obtained by the Pareto-optimal sets. Furthermore, as expected, the simulation obtained by the parameters optimised for low flow RMSE reproduces low flow accurately, while the simulation obtained by the parameters optimised for high flow RMSE is close to the high flow events. The preferred Pareto-optimal set gives the compromise simulation.

7.5.3 Uncertainty assessment

Equifinality principle of Beven and Binley (1992) is restated here that no unique optimum parameter set exists in calibration of hydrological model, but rather there will always be several different models that mimic equally well an observed natural process. In a multiobjective context, there is a multitude of parameter combinations that are equally good in reproducing the observed responses; thus multiobjective calibration accords with the equifinality principle. These equally good parameters sets are the Pareto-optimal set.

An illustration of the interpretation of Pareto-optimal sets of parameter within equifinality principle is shown in Figure 7.35. It is observed that parameter sets cover a large range of parameter values, but produce virtually equal good simulations according to a specified objective function criterion. Consequently these parameter sets produce a range of simulated hydrographs as shown in Figure 7.37. These figures illustrate that there are considerable ranges of the model simulations which are equally good according to a specified objective function criterion. In a way, this range can be interpreted as uncertainty of the model prediction corresponding to the Pareto-optimal parameter sets, and it provides additional information necessary for risk based decision making. With respect to the UNEEC method, as mentioned in Chapter 6, we could apply the UNEEC method for each solution of the Pareto-optimal sets, thus producing the several uncertainty bounds corresponding to Pareto-optimal sets. More research is anticipated in the future in this direction.

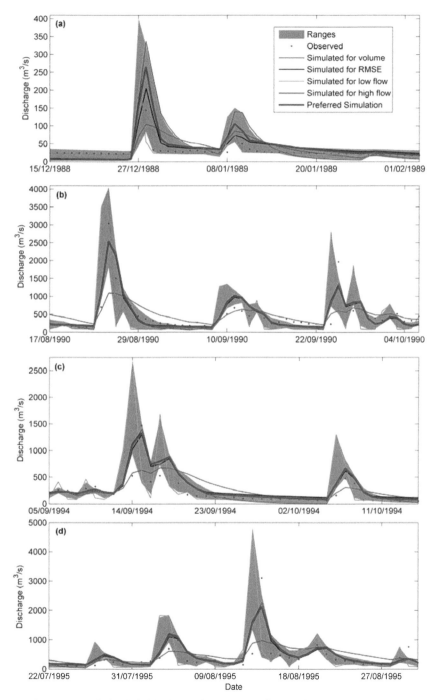

Figure 7.37. Ranges of simulated hydrographs corresponding to the Pareto parameter sets. The ranges are also compared with the solutions obtained with four single objective functions and preferred solution. (a and b) Part of the calibration period and (c and d) part of the verification period.

7.6 Conclusions

This chapter presents the application of UNEEC method to estimate uncertainty in the model prediction for the three different catchments, which have highly varied hydro-climatic characteristics. The uncertainty bound of the model prediction estimated with the UNEEC method is compared with the observed data. The results show that the percentage of the observed discharge falling within the estimated uncertainty bound is very close to the specified confidence level used to produce these bounds. It has been demonstrated that uncertainty bounds are consistent with the order of the magnitude of the model errors.

A comparative assessment of the uncertainty bounds with those obtained by other existing uncertainty methods including GLUE, the meta-Gaussian, the quantile regression indicates that the uncertainty estimation with the UNEEC method is reasonably accurate. The UNEEC method is computationally efficient and can be applied to any kind of models. This application concludes that the machine learning techniques can be used to analyse, model and predict the residual uncertainty of the optimal model output. Note that residual uncertainty considers all sources of uncertainty in aggregated form through model errors.

Chapter 8
Conclusions and
Recommendations

This research explores machine learning techniques for uncertainty analysis and prediction, particularly in rainfall-runoff modelling. This chapter presents conclusions drawn from this research and recommendations for future research and development for uncertainty analysis in the field of rainfall-runoff modelling.

8.1   Rainfall-runoff modelling and uncertainty analysis

Rainfall-runoff models are widely used in hydrology for a large range of applications and play an important role in optimal planning and management of water resources in river basins. By definition, a rainfall-runoff model is only an abstraction of a complex, non-linear, time and space varying hydrological process of reality; there are many simplifications and idealisations. These models contain parameters that cannot often be measured directly, but can only be estimated by calibration against a historical record of measured output data. The system input (forcing) data such as rainfall, temperature, etc. and observed output are often contaminated by measurement errors. Consequently predictions made by such a model are far from being perfect and uncertain, no matter how sophisticated the models are and how perfectly the models are calibrated. It is vital, therefore, that uncertainty should be recognized and properly accounted for.

Once the existence of uncertainty in a rainfall-runoff model is acknowledged, it should be managed by a proper uncertainty analysis and prediction procedures aimed eventually at reducing its impact. There a number of such procedures actively used but our analysis makes it possible to conclude that they are often based on strong assumptions and suffer from certain deficiencies. This thesis is devoted to developing new procedures for uncertainty analysis and prediction and testing them on various case studies.

It should be noted that still the practice of uncertainty analysis and use of the results of such analysis in decision making is not widespread (Pappenberger and Beven, 2006). It is not always clear for practitioners how uncertainty analysis will contribute to improved decision making. The uncertainty analysis requires careful interpretation in order to understand the meaning and significance of the results. It is through this process of scrutiny and discussion that the most useful insights for decision makers are obtained (Hall and Solomatine, 2008).

8.2 Uncertainty analysis methods

This thesis investigates a number of methods proposed in the literature to provide meaningful uncertainty bounds of the model predictions. The uncertainty analysis methods in rainfall-runoff models vary mainly in the following ways: (i) the type of rainfall-runoff models used; (ii) the source of uncertainty to be treated; (iii) the representation of uncertainty; (iv) the purpose of the uncertainty analysis; and (v) the availability of resources. Uncertainty analysis is a well-accepted procedure and has comparatively long history in physically based and conceptual modelling.

Uncertainty analysis methods can be broadly classified into six categories (see, e.g., Montanari, 2007; Shrestha and Solomatine, 2008): (i) analytical methods; (ii) approximation methods; (iii) simulation and sampling-based methods; (iv) Bayesian methods; (v) methods based on the analysis of model errors; and (vi) fuzzy set theory-based methods. Most of the existing methods analyse the uncertainty of the uncertain input variables by propagating it through the deterministic model to the outputs, and hence require the assumption of their distributions and error structures. Most of the approaches based on the analysis of the model errors require certain assumptions regarding the residuals (e.g., normality and homoscedasticity). Obviously, the relevancy and accuracy of such approaches depend on the validity of these assumptions. The fuzzy theory-based approach requires knowledge of the membership function of the quantity subject to the uncertainty, which could be very subjective. Furthermore, the majority of the uncertainty methods account for only a single source of uncertainty and ignore other sources of uncertainty explicitly. Methods based on the analysis of the model errors typically compute the uncertainty of the "optimal model" that takes into account all sources of errors without attempting to disaggregate the contribution given by their individual sources. No single method of uncertainty estimation can be claimed as being perfect in representing uncertainty. Our analysis allows us to conclude that the machine learning methods which are able to build accurate models based on data (in this case, on the data about the past model errors) have excellent potential for their use as uncertainty predictors.

8.3 Machine learning methods for uncertainty analysis

Over the last 15 years many machine learning techniques have been used to build data-driven rainfall-runoff models. There are also examples of applying these techniques as error correctors to improve the accuracy of prediction/forecasting made by process based rainfall-runoff models. Generally they are used to update the output variables by forecasting the error of the process based models. Such techniques to update the model predictions are in fact reducing the uncertainty of predictions. However these techniques do not provide explicitly the uncertainty of the model prediction in the form of prediction bounds or probability distribution function of the model output.

In this thesis we explore the possibility of using machine learning techniques to provide reasonable uncertainty estimation for the model output predicted by data-driven or process based models. We develop two methods, namely the MLUE for parametric and the UNEEC method for residual uncertainty analysis of rainfall-runoff models.

8.3.1   A method for parameter uncertainty analysis

Monte Carlo (MC) simulation is a widely used method for uncertainty analysis in rainfall-runoff modeling and allows the quantification of the model output uncertainty resulting from uncertain model parameters. It involves random sampling from the distribution of uncertain input and successive model runs until a desired statistically significant distribution of outputs is obtained. The MC based methods for uncertainty analysis of the outputs of the process models are flexible, robust, conceptually simple and straightforward; however methods of this type require a large number of samples (or model runs), and their applicability is sometimes limited to simple models. In the case of computationally intensive models, the time and resources required by these methods could be prohibitively expensive.

A number of methods have been developed to improve the efficiency of MC based uncertainty analysis methods and still these methods require a considerable number of model runs in both offline and operational mode to produce a reliable and meaningful uncertainty estimation. In this thesis we develop a method to predict parametric uncertainty of rainfall-runoff model by building machine learning models that emulate the MC uncertainty results. The proposed method is referred to as the MLUE (**M**achine **L**earning in parameter **U**ncertainty **E**stimation). The motivation to develop MLUE method is to perform fast parameter uncertainty analysis and prediction. We assume that the uncertainty of the model prediction at particular time step depends on the corresponding forcing input data and the model states (e.g., rainfall, antecedent rainfall, soil moisture etc.). We believe that uncertainty associated with prediction of hydrological variables such as runoff in similar hydrological conditions is also similar.

The MLUE method emulates the MC simulations and belongs to the class of surrogate, or meta-models. The novelty and the characteristics of the methodology are:

1. The method explicitly builds an emulator for the MC uncertainty results while other methods build an emulator for a single simulation model;

2. The MLUE emulator is based on machine learning techniques, while other techniques are Bayesian (e.g., O'Hagan, 2006), or use nonlinear differential equations (e.g., data based mechanistic model of Young (1998));

3. The method is computationally efficient and does not involve any additional runs of the process model; it can therefore be easily applied to computationally demanding process models;

4. The method can be applied easily to any Monte Carlo based uncertainty analysis methods.

The MLUE method is applied to a conceptual rainfall-runoff model for the Brue catchment in UK. The generalised likelihood uncertainty estimation method (GLUE) has been used to analyse the parameter uncertainty of the model. Machine learning methods have been applied to estimate the uncertainty results (e.g., quantile or prediction intervals) generated by the GLUE method. We have shown how domain knowledge and analytical techniques are used to select the input data for the machine learning models used in the MLUE method.

Three machine learning models, namely artificial neural networks, model trees, and locally weighted regression, are used to predict the uncertainty of the model predictions. The performance of the MLUE method is measured by its predictive capability (e.g., coefficient of correlation and root mean squared error) and the statistics of the uncertainty (e.g., the prediction intervals coverage probability and the mean prediction intervals). It is demonstrated that machine learning methods can predict the uncertainty results with reasonable accuracy. The great advantage of the MLUE method is that once the machine learning models are developed, which is done offline, it can predict the uncertainty of the model output in a fraction of second which otherwise would take several hours or day of computation time by the MC based uncertainty analysis methods. The proposed techniques could be useful in real time applications when it is impracticable to run a large number of simulations for complex hydrological models for an uncertainty analysis and when the forecast lead time is very short.

8.3.2 A method for residual uncertainty analysis

Analysis of research literature has shown that the assessment of model uncertainty of the optimal (calibrated) rainfall-runoff models has received relatively little attention. Most research typically focuses on one single source of uncertainty and the majority of the studies are oriented toward parametric uncertainty. There are many situations, however, when the contribution of the parameter uncertainty to the total uncertainty is smaller compared to the other types, for instance input (rainfall) uncertainty or structure uncertainty. The consequence of considering only parametric uncertainty is that the predictive uncertainty bounds estimated are too narrow and thus a considerable part of the observed data fall outside these bounds. Furthermore, disaggregation of the total model uncertainty into its source components is difficult, particularly in cases common to hydrology where the model is non-linear and complex, and different sources of uncertainty may interact.

Generally the analysis of uncertainty consists of propagating the uncertainty of the input and parameters (which is measured by distribution function) through the model by running it for sufficient number of times and deriving the distribution function of the model outputs. In this thesis, we present a different approach to analyzing the uncertainty. We focus on residual uncertainty which is defined as the remaining uncertainty of the optimal model. Here the model optimality is understood in the following sense: the model is calibrated, so that the model parameters and structure are such that the model error is at minimum. However, even such a model simulates or predicts the output variable with errors, so its output contains uncertainty.

We develop a novel methodology to estimate the uncertainty of the optimal model output by analyzing historical model residuals errors. This method is referred to as **UN**certainty **E**stimation based on **L**ocal **E**rrors and **C**lustering (UNEEC). The characteristics of the methodology are:

1. Residuals are used to characterize the uncertainty of the model prediction;

2. No assumptions are made about the probability distribution function of the model residuals;

3. The method uses the concept of the model optimality and does not involve any additional runs of the process model;

4. Specialized uncertainty models are built for particular areas of the state space (e.g., hydrometeorological condition). Clustering is needed to identify these areas;

5. The uncertainty models are built using machine learning techniques;

6. The method is computationally efficient and can therefore be easily applied to computationally demanding process models;

7. The method can be applied easily to any existing model no matter whether it is physically based or conceptual or data-driven.

The UNEEC method consists of the three main steps: (i) clustering the input data in order to identify the homogenous regions of the input space; (ii) estimating the probability distribution of the model residuals for the regions identified by clustering; and (iii) building the machine learning models of the probability distributions of the model error. Fuzzy clustering has been used to cluster the input data. Three machine learning models, namely artificial neural networks, model trees, locally weighted regression, are used to predict the uncertainty of the model predictions. We apply domain knowledge and analytical techniques to select the input data for the machine learning models used in the UNEEC method.

The UNEEC method is applied to rainfall-runoff models for three contrasting catchments: (i) data-driven models of the Sieve catchment, Italy; (ii) lumped process based model of the Brue catchment, UK; and (iii) lumped process based model of the Bagmati catchment, Nepal. The performance of the UNEEC method is measured by two uncertainty statistics namely the prediction intervals coverage probability and the mean prediction intervals. It has been demonstrated that uncertainty bounds estimated by the UNEEC method are consistent with the order of the magnitude of the model errors. The results show that the percentage of the observed discharge falling within the estimated uncertainty or prediction bounds is very close to the specified confidence level used to produce these bounds. In other words, the PICP values are consistently close to the desired degree of the confidence levels used to derive the prediction bounds with the reasonable width of the bounds.

We also compare the uncertainty of the model prediction with other uncertainty estimation methods, namely generalised likelihood uncertainty estimation (GLUE), meta-Gaussian, quantile regression. The comparison results show that the UNEEC method generates the consistent and interpretable uncertainty estimates, and this is an indicator that it can be a valuable tool for assessing uncertainty of various predictive models.

8.4   Multiobjective calibration and uncertainty

Practical experience with the calibration of the rainfall-runoff model suggests that a single objective function value is often inadequate to measure properly the simulation of all the important characteristics of the system that are reflected in the observations. Furthermore, recent advances in computational power and the increased availability of distributed hydrological observations have led to more complex hydrological model, often predicting multiple hydrological fluxes simultaneously. Therefore, there is an increasing interest in multiobjective calibration of rainfall-runoff model parameters.

In this research, we have applied multiobjective optimisation routine NSGA-II to calibrate the HBV rainfall-runoff model for the Bagmati catchment in Nepal. Four objective functions that emphasise different aspects of the runoff hydrograph are the volume error, root mean squared error (RMSE), RMSE of low flows, and RMSE of peak flows. We have implemented Pareto preference ordering in order to select a small number of solutions from the four dimensional Pareto-optimal solutions. The preferred Pareto-optimal set is compared with the optimal values of the parameter set for each of the single objective functions. It is observed that the results of the preferred Pareto-optimal set is a good compromise between all four objective functions and are within the best and worst solutions from the single objective function optimisations.

Multiobjective calibration also allows the quantification of the uncertainty in a form of the range of the model simulations corresponding to the Pareto-optimal sets of the parameter; and it provides additional information necessary for risk based decision making.

8.5   Summary of conclusions

The aim of this research is to develop tools and techniques for analysing, modelling and predicting uncertainty in rainfall-runoff modelling. This research explores machine learning techniques for uncertainty analysis and prediction, particularly in rainfall-runoff modelling. We develop two methods, namely the MLUE for parametric and the UNEEC method for residual uncertainty analysis of rainfall-runoff models. We apply these methods on various case studies to analyse, model and predict uncertainty in the model predictions. This research has demonstrated that machine learning methods are able to build reasonably accurate and efficient models for predicting uncertainty. We conclude that the machine learning techniques can be valuable tools for uncertainty analysis of various predictive models.

8.6   Recommendations

- In this research the MLUE method has been used to emulate the results of GLUE method. MLUE can be also applied to other MC based uncertainty analysis methods such as Markov chain Monte Carlo sampling, Latin hypercube sampling etc. Furthermore, in the MLUE method, we consider only parameter uncertainty of the process model and ignore other sources of uncertainty explicitly. The MLUE method, in principle can be also applied to other sources of uncertainty – input and structure uncertainty individually or combination of two or more sources of uncertainty.

- The UNEEC method relies on the concept of model optimality instead of equifinality. If the assumption of model optimality i.e. the existence of a single "best" model is not valid, then all of the models that are considered "good" should be considered, as is done when the concept of equifinality is adopted. This can be achieved by combining such models in an ensemble, or by generating the meta-models of uncertainty for each possible combination of the model structure and parameter set, or even involving the uncertainty associated with the input data. Consequently, instead of having a single set of uncertainty bounds for each model prediction, there will be a set of such bounds generated for each member of an ensemble. The use of such several uncertainty bounds in the decision making process would be really challenging.

- In the UNEEC method we analyse historical model residuals resulting from single objective calibration of rainfall-runoff model using Nash-Sutcliffe model efficiency criterion. It would be important to investigate the sensitivity of the uncertainty results with other objective functions (e.g., volume error, RMSE of low flows or RMSE of high flows) used in the optimization algorithm. Another interesting research is to apply the UNEEC method for multiobjective calibration results and produce the uncertainty bounds for each Pareto-optimal solution – this will be analogous to using the equifinality principle.

- We consider two types of rainfall-runoff models – data-driven and conceptual (HBV). It is recommended to test the proposed uncertainty analysis methods to physically based and other conceptual rainfall-runoff models in various case studies. Furthermore, the methodologies have considerable potential in application to other mathematical models of water based systems.

- As a machine learning method we have used ANN, model trees and locally weighted regression methods. Further studies should be aimed at testing other machine learning techniques including support vector machines, other instance based learning methods, etc.

- In multiobjective calibration we used Pareto preference ordering to select a small number of solutions from more than 400 Pareto-optimal solutions in four dimensions. The procedure to select a small number of preferred solutions from the Pareto-optimal solutions in this research consists of two independent steps – firstly generating the Pareto-optimal solutions using multiobjective optimisation algorithm (such as NSGA-II) and secondly using preference ordering. It would be worthwhile to apply a multiobjective optimisation algorithm based on the preference ordering that is supposed to be more effective in achieving a better grading of a set of solutions to a problem that consists of many objective functions.

- Although the concept of model uncertainty is well recognized in the research community, the practice of uncertainty analysis and use of the results of such analysis in the decision making is not widespread. It is not always clear how uncertainty analysis will contribute to improving decision making. One of the most important challenges is to communicate effectively to the decision maker, the insights and the advantages an analysis of uncertainty provides and to convert uncertain results to into simple message understood by the general public. The further research should be fostered into incorporating the uncertainty analysis results in risk based decision making.

ABBREVIATIONS

ACCO	: Adaptive Clustering Coverage
AMI	: Average Mutual Information
ANN	: Artificial Neural Network
cdf	: Cumulative Distribution Function
CoC	: Coefficient of Correlation
CoE	: Nash-Sutcliffe Model Efficiency
FCM	: Fuzzy C-means
GLUE	: Generalised Likelihood Uncertainty Estimation
HBV	: Hydrologiska Byråns Vattenbalansavdelning (Hydrological Bureau Water balance section)
LHS	: Latin Hypercube Sampling
LWR	: Locally Weighted Regression
MC	: Monte Carlo
MCMC	: Markov Chain Monte Carlo
MLP	: Multi Layer Perceptron
MLUE	: Machine Learning in Parameter Uncertainty Estimation
MPI	: Mean Prediction Interval
MT	: Model Trees
NQT	: Normal Quantile Transform
NSGA	: Non-dominated Sorting Genetic Algorithm
pdf	: Probability Distribution Function
PI	: Prediction Interval
PICP	: Prediction Interval Coverage Probability
QR	: Quantile Regression
RMSE	: Root Mean Squared Error
SNR	: Signal-to-Noise Ratio
UNEEC	: Uncertainty Estimation based on Local Errors and Clustering

NOTATIONS

y_t^p : pth quantile of the model output

μ : Fuzzy membership function, [0,1]

ε : Model error

θ : Model parameters

α : Significant level used to derive prediction intervals, (0, 1)

C   : Cluster label

C   : Number of cluster

D   : Input data {X, **y**}

d(.) : Distance function

e   : Quantile of error

ec   : Quantile of error for cluster

J   : Objective function

m   : Fuzzy exponential coefficient, [1,∞]

M   : Primary model

n   : Number of data

O   : Order of computational complexity

p   : Number of variables

PI^L : Lower prediction interval

PI^U : Upper prediction interval

PL^L : Lower prediction limit

PL^U : Upper prediction limit

Q   : Discharge (L^3/T)

R   : Rainfall (L/T)

RE : Effective rainfall (L/T)

s   : Number of simulations

P   : Partition matrix

U   : Uncertainty model

v   : Centre of clusters

w   : Weight vector

x, X **:** Input vector, matrix

y   : Observed or recorded or target output

$\hat{y}$: Model output

REFERENCES

Abbott, M.B. (1991). *Hydroinformatics: Information technology and the aquatic environment.* Avebury Technical, Aldershot, UK.

Abbott, M.B., Bathurst, J.C., Cunge, J.A., O'Connell, P.E. and Rasmussen, J. (1986a). An introduction to the European Hydrological System - System Hydrologique Europeen,"SHE", 1: History and philosophy of a physically-based, distributed modelling system. *Journal of Hydrology*, 87, pp. 45-59.

Abbott, M.B., Bathurst, J.C., Cunge, J.A., O'Connell, P.E. and Rasmussen, J. (1986b). An introduction to the European Hydrological System - Systeme Hydrologique Europeen, "SHE", 2: Structure of a physically-based, distributed modelling system. *Journal of Hydrology*, 87, pp. 61-77.

Abebe, A.J. (2004). *Information theory and artificial intelligence to manage uncertainty in hydrodynamic and hydrological models.* PhD Thesis, UNESCO-IHE Institute for Water Education, Delft, The Netherlands.

Abebe, A.J., Guinot, V. and Solomatine, D.P. (2000). Fuzzy alpha-cut vs. Monte Carlo techniques in assessing uncertainty in model parameters, *Proc. of 4th International Conference on Hydroinformatics*, Iowa city, USA.

Abebe, A.J. and Price, R.K. (2003). Managing uncertainty in hydrological models using complementary models. *Hydrological Sciences Journal*, 48(5), pp. 679-692.

Abrahart, R., Kneale, P. and See, L. (2004). *Neural networks for hydrological modelling.* A.A. Balkema Publishers, Leiden, The Netherlands.

Abrahart, R.J. and See, L. (2000). Comparing neural network and autoregressive moving average techniques for the provision of continuous river flow forecasts in two contrasting catchments. *Hydrological Processes*, 14, pp. 2157-2172.

Aha, D., Kibler, D. and Albert, M. (1991). Instance-based learning algorithms. *Machine Learning*, 6, pp. 37-66.

Ajami, N.K., Duan, Q. and Sorooshian, S. (2007). An integrated hydrologic Bayesian multimodel combination framework: Confronting input, parameter, and model structural uncertainty in hydrologic prediction. *Water Resources Research*, 43, W01403, doi:10.1029/2005WR004745.

Allen, R.G., Pereira, L.S., Raes, D. and Smith, M. (1998). *Crop evapotranspiration - guidelines for computing crop water requirements.* Irrigation and Drainage Paper No. 56, FAO, Rome. Available at: http://www.fao.org/docrep/X0490E/x0490e00.htm.

Amorocho, J. and Espildora, B. (1973). Entropy in the assessment of uncertainty in hydrologic systems and models. *Water Resources Research*, 9(6), pp. 1511-1522.

Ang, A.H.S. and Tang, W.H. (1975). *Probability concepts in engineering planning and design: Volume 1–basic principles.* John Wiley & Sons, New York, NY, USA.

ASCE (2000). Artificial neural networks in hydrology. I: Preliminary concepts. *Journal of Hydrologic Engineering*, 5(2), pp. 115-123.

Ayyub, B. and Chao, R. (1998). Uncertainty modeling in civil engineering with structural and reliability applications. In: B. Ayyub (ed.), *Uncertainty modeling and analysis in civil engineering*, CRC Press, Boca Raton, FL, USA, pp. 3-32.

Ballio, F. and Guadagnini, A. (2004). Convergence assessment of numerical Monte Carlo simulations in groundwater hydrology. *Water Resources Research*, 40(4), W04603, doi: 10.1029/2003WR002876.

Bardossy, A., Bogardi, I. and Duckstein, L. (1990). Fuzzy regression in hydrology. *Water Resources Research*, 26(7), pp. 1497-1508.

Bárdossy, A. and Duckstein, L. (1995). *Fuzzy rule-based modeling with applications to geophysical, biological, and engineering systems*. CRC Press, Boca Raton, FL, USA.

Bell, V.A. and Moore, R.J. (2000). The sensitivity of catchment runoff models to rainfall data at different spatial scales. *Hydrology and Earth System Sciences*, 4(4), pp. 653-667.

Bensaid, A.M., Hall, L.O., Bezdek, J.C., Clarke, L.P., Silbiger, M.L., Arrington, J.A. and Murtagh, R.F. (1996). Validity-guided (re)clustering with applications to image segmentation. *IEEE Transactions on Fuzzy Systems*, 4(2), pp. 112-123.

Bergström, S. (1976). *Development and application of a conceptual runoff model for Scandinavian catchments*. SMHI Reports RHO, No. 7, Norrköping, Sweden.

Bergström, S. and Forsman, A. (1973). Development of a conceptual deterministic rainfall-runoff model. *Nordic Hydrology*, 4, pp. 147-170.

Beven, K. (1989). Changing ideas in hydrology - the case of physically based models. *Journal of Hydrology*, 105, pp. 157-172.

Beven, K. (2001). How far can we go in distributed hydrological modelling? *Hydrology and Earth System Sciences*, 5(1), pp. 1-12.

Beven, K. (2006). A manifesto for the equifinality thesis. *Journal of Hydrology*, 320(1), pp. 18-36.

Beven, K. and Binley, A. (1992). The future of distributed models: Model calibration and uncertainty prediction. *Hydrological Processes*, 6, pp. 279-298.

Beven, K. and Freer, J. (2001). Equifinality, data assimilation, and uncertainty estimation in mechanistic modelling of complex environmental systems using the GLUE methodology. *Journal of Hydrology*, 249, pp. 11-29.

Beven, K. and Kirkby, M. (1979). A physically based, variable contributing area model of basin hydrology. *Hydrological Sciences Bulletin*, 24(1), pp. 43-69.

Bezdek, J.C. (1981). *Pattern recognition with fuzzy objective function algorithms*. Kluwer Academic Publishers, Norwell, MA, USA.

Blasone, R., Vrugt, J., Madsen, H., Rosbjerg, D., Robinson, B. and Zyvoloski, G. (2008). Generalized likelihood uncertainty estimation (GLUE) using adaptive Markov Chain Monte Carlo sampling. *Advances in Water Resources*, 31, pp. 630-648.

Bowden, G.J., Dandy, G.C. and Maier, H.R. (2005). Input determination for neural network models in water resources applications. Part 1—background and methodology. *Journal of Hydrology*, 301, pp. 75-92.

Box, G. and Cox, D. (1964). An analysis of transformations. *Journal of the Royal Statistical Society. Series B (Methodological)*, pp. 211-252.

Box, G.E.P. and Jenkins, G.M. (1970). *Time series analysis*. Holden-Day, San Francisco, CA, USA.

Brath, A., Montanari, A. and Toth, E. (2002). Neural networks and non-parametric methods for improving real-time flood forecasting through conceptual hydrological models. *Hydrology and Earth System Sciences*, 6(4), pp. 627-639.

Braun, L.N. and Renner, C.B. (1992). Application of a conceptual runoff model in different physiographic regions of Switzerland. *Hydrological Sciences Journal*, 37(3), pp. 217-232.

Bray, M. and Han, D. (2004). Identification of support vector machines for runoff modelling. *Journal of Hydroinformatics*, 6, pp. 265-280.

Breiman, L. (1996). Bagging predictors. *Machine Learning*, 24(2), pp. 123-140.

Breiman, L. (2001). Statistical modeling : The two cultures. *Statistical Science*, 16(3), pp. 199-231.

Breiman, L., Friedman, J., Olshen, R. and Stone, C. (1984). *Classification and regression trees*. Chapman & Hall/CRC, Boca Raton, FL, USA.

Brown, J.D. and Heuvelink, G.B.M. (2005). Assessing uncertainty propagation through physically based models of soil water flow and solute transport. In: M. Anderson (ed.), *Encyclopedia of Hydrological Sciences*, John Wiley & Sons, New York, NY, USA, pp. 1181-1195.

Burges, S. and Lettenmaier, D. (1975). Probabilistic methods in stream quality management. *Water Resources Bulletin*, 11(1), pp. 115-130.

Burnash, R.J.C., Ferral, R.L. and McGuire, R.A. (1973). *A generalized streamflow simulation system, conceptual modeling for digital computers*. Report by the Joint Federal State River Forecasting Centre, Sacramento, CA, USA.

Butts, M.B., Payne, J.T., Kristensen, M. and Madsen, H. (2004). An evaluation of the impact of model structure on hydrological modelling uncertainty for streamflow simulation. *Journal of Hydrology*, 298, pp. 222-241.

Cawley, G.C., Talbot, N.L.C., Foxall, R.J., Dorling, S.R. and Mandic, D.P. (2004). Heteroscedastic kernel ridge regression. *Neurocomputing*, 57, pp. 105-124.

Chakraborty, K., Mehrotra, K., Mohan, C. and Ranka, S. (1992). Forecasting the behavior of multivariate time series using neural networks. *Neural Networks*, 5(6), pp. 961-970.

Chalise, S.R., Kansakar, S.R., Rees, G., Croker, K. and Zaidman, M. (2003). Management of water resources and low flow estimation for the Himalayan basins of Nepal. *Journal of Hydrology*, 282(1-4), pp. 25-35.

Chatfield, C. (2000). *Time-Series forecasting*. Chapman & Hall/CRC, Boca Raton, FL, USA.

Chow, V. (1964). *Handbook of applied hydrology*. McGraw-Hill, New York, NY, USA.

Chow, V., Maidment, D. and Mays, L. (1988). *Applied hydrology*. McGraw-Hill, New York, NY, USA.

Clarke, R. (1973). A review of some mathematical models used in hydrology, with observations on their calibration and use. *Journal of hydrology*, 19(1), pp. 1-20.

Cleveland, W.S. and Loader, C. (1994). *Smoothing by local regression: principles and methods*, AT and T Bell Laboratories, Statistics Department, Murray Hill, NJ, USA.

Corzo, G., Solomatine, D., Hidayat, de Wit, M., Werner, M., Uhlenbrook, S. and Price, R.K. (2009). Combining semi-distributed process-based and data-driven models in flow simulation: a case study of the Meuse river basin. *Hydrology and Earth System Sciences Discussions*, 6(1), pp. 729-766.

Crawford, N. and Linsley, R. (1966). *Digital simulation in hydrology: Stanford Watershed Model IV*. Technical Report No. 39, Department of Civil Engineering, Stanford University, Stanford, CA, USA.

Das, I. (1999). A preference ordering among various Pareto optimal alternatives. *Structural Optimization*, 18(1), pp. 30-35.

David, J.P. and Blockley, D.I. (1996). On modelling uncertainty. In: A. Muller (ed.), *Proc. of The 2nd International Conference on Hydroinformatics*, Zurich, Switzerland, Balkema, Rotterdam, pp. 485-491.

Dawson, C.W. and Wilby, R. (1998). An artificial neural network approach to rainfall-runoff modelling. *Hydrological Science Journal*, 43(1), pp. 47-66.

Deb, K., Pratap, A., Agarwal, S. and Meyarivan, T. (2002). A fast and elitist multiobjective genetic algorithm: NSGA-II. *IEEE Transactions on Evolutionary Computation*, 6(2), pp. 182-197.

DHM (1998). *Hydrological records of Nepal, stream flow summary*, Department of Hydrology and Meteorology, Kathmandu, Nepal.

Dibike, Y.B., Velickov, S., Solomatine, D. and Abbott, M.B. (2001). Model induction with support vector machines: Introduction and applications. *Journal of Computing in Civil Engineering*, 15(3), pp. 208-216.

Duan, Q., Sorooshian, S. and Gupta, V. (1992). Effective and efficient global optimization for conceptual rainfall-runoff models. *Water Resources Research*, 28(4), pp. 1015-1031.

Dunn, J. (1973). A fuzzy relative of the Isodata process and its use in detecting compact, well-separated clusters. *Journal of Cybernetics*, 3(3), pp. 32-57.

Engeland, K., Xu, C.-Y. and Gottschalk, L. (2005). Assessing uncertainties in a conceptual water balance model using Bayesian methodology. *Hydrological Sciences Journal*, 50(1), pp. 45-63.

Fleming, G. (1975). *Computer simulation techniques in hydrology*. Elsevier, New York, NY, USA.

Franchini, M. (1996). Use of genetic algorithm combined with a local search method for the automatic calibration of conceptual rainfall-runoff models. *Hydrological Science Journal*, 41(1), pp. 21-39.

Freer, J., Beven, K. and Ambroise, B. (1996). Bayesian estimation of uncertainty in runoff prediction and the value of data: An application of the GLUE approach. *Water Resources Research*, 32(7), pp. 2161-2173.

Freund, Y. and Schapire, R.E. (1996). Experiments with a new boosting algorithm. In: L. Saitta (ed.), *Proc. of 13th International Conference on Machine Learning*, Bari, Italy, Morgan Kaufmann, San Francisco, CA, USA, pp. 148-156.

Frieser, B.I., Vrijling, J.K. and Jonkman, S.N. (2005). Probabilistic evacuation decision model for river floods in the Netherlands, *Proc. of 9th International Symposium on Stochastic Hydraulics*, Nijmegen, The Netherlands, IAHR.

Geman, S., Bienenstock, E. and Doursat, R. (1992). Neural networks and the bias/variance dilemma. *Neural Computation*, 4(1), pp. 1-58.

Georgakakos, K.P., Seo, D.-J., Gupta, H., Schaake, J. and Butts, M.B. (2004). Towards the characterization of streamflow simulation uncertainty through multimodel ensembles. *Journal of Hydrology*, 298, pp. 222-241.

Goldstein, M. and Rougier, J.C. (2009). Reified Bayesian modelling and inference for physical systems. *Journal of Statistical Planning and Inference*, 139, pp. 1221-1239.

Gouldby, B. and Samuels, P. (2005). *Language of risk . Project definitions*. **FLOOD***site* Consortium Report T32-04-01. Available at: www.floodsite.net.

Govindaraju, R.S. and Rao, A.R. (2000). *Artificial neural networks in hydrology*. Kluwer Academic Publishers, Dordrecht, The Netherlands.

Guinot, V. and Gourbesville, P. (2003). Calibration of physically based models: back to basics? *Journal of Hydroinformatics*, 5(4), pp. 233-244.

Gupta, H., Sorooshian, S. and Yapo, P. (1998). Toward improved calibration of hydrologic models: Multiple and noncommensurable measures of information. *Water Resources Research*, 34(4), pp. 751-763.

Gupta, H.V., Beven, K.J. and Wagener, T. (2005). Model calibration and uncertainty estimation. In: M. Andersen (ed.), *Encyclopedia of Hydrological Sciences*, John Wiley & Sons, New York, NY, USA, pp. 2015-2031.

Guyon, I. and Elisseeff, A. (2003). An introduction to variable and feature selection. *Journal of Machine Learning Research*, 3(7-8), pp. 1157-1182.

Haario, H., Laine, M., Mira, A. and Saksman, E. (2006). DRAM: efficient adaptive MCMC. *Statistics and Computing*, 16(339-354), doi: 10.1007/s11222-006-9438-0.

Hall, J. (1999). *Uncertainty management for coastal defence systems*. PhD Thesis, Department of Civil Engineering, Bristol University, Bristol, UK.

Hall, J. (2003). Handling uncertainty in the hydroinformatics process. *Journal of Hydroinformatics*, 05(4), pp. 215-232.

Hall, J. and Anderson, M. (2002). Handling uncertainty in extreme or unrepeatable hydrological processes—the need for an alternative paradigm. *Hydrological Processes*, 16, pp. 1867-1870, doi: 10.1002/hyp.5026.

Hall, J. and Solomatine, D. (2008). A framework for uncertainty analysis in flood risk management decisions. *Intl. J. River Basin Management*, 6(2), pp. 85-98.

Harnett, D.L. and Murphy, J.L. (1980). *Introductory statistical analysis*. Addison-Wesley Publishing Company, Reading, MA, USA.

Harr, M.E. (1989). Probabilistic estimates for multivariate analyses. *Applied Mathematical Modelling*, 13(5), pp. 313-318.

Hastie, T., Tibshirani, R. and Friedman, J. (2001). *The elements of statistical learning*. Springer, New York, NY, USA.

Hastings, W.K. (1970). Monte Carlo sampling methods using Markov chains and their applications. *Biometrika*, 57(1), pp. 97-109.

Haykin, S. (1999). *Neural networks: A comprehensive foundation*. Prentice Hall, Upper Saddle River, NJ, USA.

Helton, J.C. and Davis, F.J. (2003). Latin hypercube sampling and the propagation of uncertainty in analyses of complex systems. *Reliability Engineering & System Safety*, 81(1), pp. 23-69.

Hsu, K., Gupta, H.V. and Sorooshian, S. (1995). Artificial neural network modelling of the rainfall-runoff process. *Water Resources Research*, 31(10), pp. 2517-2530.

IPCC (2005). *Guidance notes for lead authors of the IPCC fourth assessment report on addressing uncertainties*, Intergovernmental Panel on Climate Change. Available at: http://www.ipcc.ch/pdf/assessment-report/ar4/wg1/ar4-uncertaintyguidancenote.pdf.

Jacobs, R., Jordan, M., Nowlan, S. and Hinton, G. (1991). Adaptive mixtures of local experts. *Neural Computation*, 3(1), pp. 79-87.

Johnston, P. and Pilgrim, D. (1976). Parameter optimization for watershed models. *Water Resources Research*, 12(3), pp. 477-486.

Kaufmann, A. and Gupta, M. (1991). *Introduction to fuzzy arithmetic: theory and applications*. Van Nostrand Reinhold Company.

Kavetski, D., Kuczera, G. and Franks, S.W. (2006). Bayesian analysis of input uncertainty in hydrological modeling: 1. Theory. *Water Resources Research*, 42, W03407, doi: 10.1029/2005WR004368.

Kelly, K.S. and Krzysztofowicz, R. (1997). A bivariate meta-Gaussian density for use in hydrology. *Stochastic Hydrology and Hydraulics*, 11(1), pp. 17-31.

Kelly, K.S. and Krzysztofowicz, R. (2000). Precipitation uncertainty processor for probabilistic river stage forecasting. *Water Resources Research*, 36(9), pp. 2643-2653.

Kennedy, M.C. and O'Hagan, A. (2001). Bayesian calibration of computer models. *Journal of the Royal Statistical Society, Series B*, 63(3), pp. 425-464.

Khu, S. and Madsen, H. (2005). Multiobjective calibration with Pareto preference ordering: An application to rainfall-runoff model calibration. *Water Resources Research*, 41, W03004, doi: 10.1029/2004WR003041.

Khu, S.T. and Werner, M.G.F. (2003). Reduction of Monte-Carlo simulation runs for uncertainty estimation in hydrological modelling. *Hydrology and Earth System Sciences*, 7(5), pp. 680-692.

Kitanidis, P.K. and Bras, R.L. (1980). Real-Time forecasting with a conceptual hydrologic model: Analysis of uncertainty. *Water Resources Research*, 16(6), pp. 1025-1033.

Klir, G. and Folger, T. (1988). *Fuzzy sets, uncertainty, and information.* Prentice Hall, Englewood Cliffs, NJ, USA.

Klir, G.J. and Wierman, M.J. (1998). *Uncertainty-based information.* Physica-Verlag, Heidelberg, Germany.

Koenker, R. and Bassett, G. (1978). Regression quantiles. *Econometrica*, 46(1), pp. 33-50.

Koller, D. and Sahami, M. (1996). Toward optimal feature selection. In: L. Saitta (ed.), *Proc. of 13th International Conference on Machine Learning*, Bari, Italy, Morgan Kaufmann, San Francisco, CA, USA, pp. 284-292.

Kottegoda, N. (1980). *Stochastic water resources technology.* Macmillan, New York, NY, USA.

Krause, P. and Clark, D. (1993). *Representing uncertain knowledge: An artificial intelligence approach.* Kluwer Academic Publishers, Norwell, MA, USA.

Krzysztofowicz, R. (1999). Bayesian theory of probabilistic forecasting via deterministic hydrologic model. *Water Resources Research*, 35(9), pp. 2739-2750.

Krzysztofowicz, R. (2001a). The case for probabilistic forecasting in hydrology. *Journal of Hydrology*, 249, pp. 2-9.

Krzysztofowicz, R. (2001b). Integrator of uncertainties for probabilistic river stage forecasting: precipitation-dependent model. *Journal of Hydrology*, 249, pp. 69-55.

Krzysztofowicz, R. and Kelly, K.S. (2000). Hydrologic uncertainty processor for probabilistic river stage forecasting. *Water Resources Research*, 36(11), pp. 3265-3277.

Kuczera, G. and Parent, E. (1998). Monte Carlo assessment of parameter uncertainty in conceptual catchment models: the Metropolis algorithm. *Journal of Hydrology*, 211, pp. 69-85.

Kundzewicz, Z. (1995). Hydrological uncertainty in perspective. In: Z. Kundzewicz (ed.), *New Uncertainty Concepts in Hydrology and Water Resources*, University Press, Cambridge, UK, pp. 3-10.

Kunstmann, H., Kinzelbach, W. and Siegfried, T. (2002). Conditional first-order second moment method and its application to the quantification of uncertainty in groundwater modeling. *Water Resources Research*, 38(4), pp. 1035, doi: 10.1029/2000WR000022.

Langley, R.S. (2000). Unified approach to probabilistic and possibilistic analysis of uncertain system. *Journal of Engineering Mechanics*, 126(11), pp. 1163-1172.

LeBaron, B. and Weigend, A.S. (1994). Evaluating neural network predictors by bootstrapping, *Proc. of The International Conference on Neural Information Processing (ICONIP'94)*, Seoul, Korea.

Li, G., Azarm, S., Farhang-Mehr, A. and Diaz, A. (2006). Approximation of multiresponse deterministic engineering simulations: a dependent metamodeling approach. *Structural and Multidisciplinary Optimization*, 31(4), pp. 260-269.

Lilliefors, H.W. (1967). On the Kolmogorov-Smirnov test for normality with mean and variance unknown. *Journal of the American Statistical Association*, 62(318), pp. 399-402.

Lindström, G., Johansson, B., Persson, M., Gardelin, M. and Bergström, S. (1997). Development and test of the distributed HBV-96 hydrological model. *Journal of Hydrology*, 201, pp. 272-228.

Liong, S., Gautam, T., Khu, S., Babovic, V., Keijzer, M. and Muttil, N. (2002). Genetic programming: A new paradigm in rainfall runoff modeling. *Journal of the American Water Resources Association*, 38(3), pp. 705-718.

Liu, Y. and Gupta, H.V. (2007). Uncertainty in hydrologic modeling: Toward an integrated data assimilation framework. *Water Resources Research*, 43, W07401, doi: 10.1029/2006WR005756.

Madsen, H. (2000). Automatic calibration of a conceptual rainfall–runoff model using multiple objectives. *Journal of Hydrology*, 235, pp. 276-288.

Maier, H.R. and Dandy, G.C. (2000). Neural networks for the prediction and forecasting of water resources variables: a review of modeling issues and applications. *Environmental Modelling and Software*, 15, pp. 101-124.

Mantovan, P. and Todini, E. (2006). Hydrological forecasting uncertainty assessment: Incoherence of the GLUE methodology. *Journal of Hydrology*, 330(1-2), pp. 368-381.

Markus, M., Tsai, C.W.-S. and Demissie, M. (2003). Uncertainty of weekly nitrate-nitrogen forecasts using artificial neural networks. *Journal of Environmental Engineering*, 129(3), pp. 267-274, doi: 10.1061/(ASCE)0733-9372(2003)129:3(267).

Maskey, S. (2004). *Modelling uncertainty in flood forecasting systems*. PhD Thesis, UNESCO-IHE Institute for Water Education, Delft, The Netherlands.

Maskey, S. and Guinot, V. (2003). Improved first order second moment method for uncertainty estimation in flood forecasting. *Hydrological Sciences Journal*, 48(2), pp. 183-196.

Maskey, S., Guinot, V. and Price, R.K. (2004). Treatment of precipitation uncertainty in rainfall-runoff modelling: a fuzzy set approach. *Advances in Water Resources*, 27, pp. 889-898.

McCulloch, W. and Pitts, W. (1943). A logical calculus of the ideas immanent in nervous activity. *Bulletin of Mathematical Biophysics*, 5(4), pp. 115-133.

McKay, M.D. (1988). Sensitivity and uncertainty analysis using a statistical sample of input values. In: Y. Ronen (ed.), *Uncertainty Analysis*, CRC Press, Boca Raton, FL, USA, pp. 145-186.

McKay, M.D., Beckman, R.J. and Conover, W.J. (1979). A comparison of three methods for selecting values of input variables in the analysis of output from a computer code. *Technometrics*, 21(2), pp. 239-245.

Melching, C.S. (1992). An improved first-order reliability approach for assessing uncertainties in hydrological modelling. *Journal of Hydrology*, 132, pp. 157-177.

Melching, C.S. (1995). Reliability estimation. In: V.P. Singh (ed.), *Computer Models of Watershed Hydrology*, Water Resources Publications, Highlands Ranch, CO, USA, pp. 69-118.

Metropolis, N., Rosenbluth, A.W., Rosenbluth, M.N. and Teller, A.H. (1953). Equations of State Calculations by Fast Computing Machines. *The Journal of Chemical Physics*, 26(6), pp. 1087-1092.

Minns, A.W. and Hall, M.J. (1996). Artificial neural networks as rainfall-runoff models. *Hydrological science Journal*, 41, pp. 399-417.

Mitchell, T.M. (1997). *Machine learning*. McGraw-Hill, Singapore.

Montanari, A. (2005). Large sample behaviors of the generalized likelihood uncertainty estimation (GLUE) in assessing the uncertainty of rainfall-runoff simulations. *Water Resources Research*, 41, W08406, doi: 10.1029/2004WR003826.

Montanari, A. (2007). What do we mean by uncertainty? The need for a consistent wording about uncertainty assessment in hydrology. *Hydrological Processes*, 21(6), pp. 841-845.

Montanari, A. and Brath, A. (2004). A stochastic approach for assessing the uncertainty of rainfall-runoff simulations. *Water Resources Research*, 40, W01106, doi: 10.1029/ 2003WR002540.

Montanari, A. and Grossi, G. (2008). Estimating the uncertainty of hydrological forecasts: A statistical approach. *Water Resources Research*, 44, W00B08, doi: 10.1029/2008WR006897.

Moore, B. (2002). Special Issue: HYREX: The hydrological radar experiment *Hydrology and Earth System Sciences*, 4(4), pp. 521-522.

Moradkhani, H. and Sorooshian, S. (2008). General review of rainfall-runoff modeling: Model calibration, data assimilation, and uncertainty analysis. In: S. Sorooshian et al. (eds.), *Hydrological Modelling and the Water Cycle*, Springer, Heidelberg, Germany, pp. 1-24.

Nash, J.E. and Sutcliffe, J.V. (1970). River flow forecasting through conceptual models Part 1- A Discussion Principles. *Journal of Hydrology*, 10, pp. 282-290.

Nielsen, S. and Hansen, E. (1973). Numerical simulation of the rainfall-runoff process on a daily basis. *Nordic Hydrology*, 4(3), pp. 171–190.

O'Hagan, A. (2006). Bayesian analysis of computer code outputs: A tutorial. *Reliability Engineering and System Safety*, 91(10-11), pp. 1290-1300.

Pappenberger, F. and Beven, K.J. (2006). Ignorance is bliss: Or seven reasons not to use uncertainty analysis. *Water Resources Research*, 42, W05302, doi: 10.1029/2005WR004820.

Pappenberger, F., Harvey, H., Beven, K., Hall, J. and Meadowcroft, I. (2006). Decision tree for choosing an uncertainty analysis methodology: a wiki experiment http://www.floodrisknet.org.uk/methods http://www.floodrisk.net. *Hydrological Processes*, 20, pp. 3793-3798.

Price, R.K. (2002). Hydroinformatics - Lecture 1, *Proc. of RBM 2002: Advanced study course on river basin modelling for flood risk mitigation*, The University of Birmingham, UK, pp. 5.1-5.19.

Price, R.K. (2006). The growth and significance of hydroinformatics. In: D.W. Knight and A.Y. Shamseldin (eds.), *River Basin Modelling for Flood Risk Mitigation*, Taylor & Francis, London, UK.

Principe, J., Euliano, N. and Lefebvre, W. (1999). *Neural and adaptive systems: Fundamentals through simulations with CD-ROM*. John Wiley & Sons, New York, NY, USA.

Protopapas, A.L. and Bras, R.L. (1990). Uncertainty propagation with numerical models for flow and transport in the unsaturated zone. *Water Resources Research*, 26(10), pp. 2463-2474.

Quinlan, J.R. (1992). Learning with continuous classes, *Proc. of The 5th Australian Joint Conference on AI*, World Scientific, Singapore, pp. 343-348.

Refsgaard, J.C. (1996). Terminology, modelling protocol and classification of hydrological model codes. In: M.B. Abbott and J.C. Refsgaard (eds.), *Distributed Hydrological Modelling*, Kluwer Academic Publishers, Dordrecht, The Netherlands, pp. 17-39.

Refsgaard, J.C. and Storm, B. (eds.) (1996). *Distributed hydrological modelling*. Kluwer Academic Publishers, Dordrecht, The Netherlands.

Rosenblueth, E. (1975). Point estimates for probability moments. *Proceedings of the National Academy of Sciences of the United State of America*, 72(10), pp. 3812-3814.

Rosenblueth, E. (1981). Two-point estimates in probabilities. *Applied Mathematical Modelling*, 5, pp. 329-335.

Ross, T.J. (1995). *Fuzzy logic with engineering applications*. McGraw-Hill, New York, NY, USA.

Rumelhart, D., Hintont, G. and Williams, R. (1986). Learning representations by back-propagating errors. *Nature*, 323(6088), pp. 533-536.

Savenije, H.H.G. (2001). Equifinality, a blessing in disguise. *Hydrological Processes*, 15, pp. 2835-2838, doi: 10.1002/hyp.494.

Schoups, G., Van de Giesen, N. and Savenije, H. (2008). Model complexity control for hydrologic prediction. *Water Resources Research*, 44, W00B03, doi: 10.1029/2008WR006836.

Scott, D. (1992). *Multivariate density estimation*. John Wiley & Sons, New York, NY, USA.

Seibert, J. (1997). Estimation of parameter uncertainty in the HBV model. *Nordic Hydrology*, 28(4), pp. 247-262.

Shafer, G. (1976). *Mathematical theory of evidence* Princeton University Press, Princeton, NJ, USA.

Shamseldin, A.Y. and O'Connor, K.M. (2001). A non-linear neural network technique for updating river flow forecasts. *Hydrology and Earth System Sciences*, 5(4), pp. 577-597.

Shannon, C.E. (1948). A mathematical theory of communication. *The Bell System Technical Journal*, 27, pp. 379-423, 623-656.

Sharma, A. (2000). Seasonal to interannual rainfall probabilistic forecasts for improved water supply management: Part 3—A nonparametric probabilistic forecast model. *Journal of Hydrology*, 239(1-4), pp. 249-258.

Sherman, L. (1932). Streamflow from rainfall by the unit-graph method. *Engineering News Record*, 108, pp. 501-505.

Shrestha, D.L., Kayastha, N. and Solomatine, D. (2009). A novel approach to parameter uncertainty analysis of hydrological models using neural networks. *Hydrology and Earth System Sciences*, 13, pp. 1235-1248.

Shrestha, D.L., Rodriguez, J., Price, R.K. and Solomatine, D. (2006). Assessing model prediction limits using fuzzy clustering and machine learning, *Proc. of 7th International Conference on Hydroinformatics*, Nice, Research Publishing Services.

Shrestha, D.L. and Solomatine, D. (2006a). Machine learning approaches for estimation of prediction interval for the model output. *Neural Networks*, 19(2), pp. 225-235.

Shrestha, D.L. and Solomatine, D. (2008). Data-driven approaches for estimating uncertainty in rainfall-runoff modelling. *Intl. J. River Basin Management*, 6(2), pp. 109-122.

Shrestha, D.L. and Solomatine, D.P. (2005). Quantifying Uncertainty of Flood Forecasting Using Data Driven Modeling, *Proc. of 29th IAHR Congress*, Seoul, South Korea, IAHR, pp. 715-726.

Shrestha, D.L. and Solomatine, D.P. (2006b). Experiments with AdaBoost.RT, an improved boosting scheme for regression. *Neural Computation*, 18(7), pp. 1678-1710.

Singh, V. and Rajagopal, A. (1987). Some recent advances in the application of the principle of maximum entropy(POME) in hydrology, *Proc. of Water for the Future: Hydrology in Perspective. IAHS Publication*, Rome.

Singh, V. and Woolhiser, D. (2002). Mathematical modeling of watershed hydrology. *Journal of Hydrologic Engineering*, 7(4), pp. 270-292.

Singh, V.P. (1995a). *Computer models of watershed hydrology*. Water Resources Publication, Highlands Ranch, CO, USA.

Singh, V.P. (1995b). Watershed modeling. In: V.P. Singh (ed.), *Computer Models of Watershed Hydrology*, Water Resources Publication, Highlands Ranch, CO, USA, pp. 1-22.

Singh, V.P. (1997). The use of entropy in hydrology and water resources. *Hydrological Processes*, 11, pp. 587-626.

Singh, V.P. (2000). The entropy theory as a tool for modelling and decision-making in environmental and water resources. *Water SA* 26(1), pp. 1-12.

Sluijs, J.P.v.d., Risbey, J.S., Kloprogge, P., Ravetz, J.R., Funtowicz, S.O., Quintana, S.C., Pereira, A.G., Marchi, B.D., Petersen, A.C., Janssen, P.H.M., Hoppe, R. and Huijs, S.W.F. (2003). *RIVM/MNP guidance for uncertainty assessment and communication: Detailed guidance*, Utrecht University, Utrecht. Available at: http://www.nusap.net/downloads/detailedguidance.pdf.

Solomatine, D. (2002). Data-driven modelling: paradigm, methods, exercises, *Proc. of 5th International Conference on Hydroinformatics*, Cardiff, UK.

Solomatine, D. (2005). Data-driven modeling and computational intelligence methods in hydrology. In: M. Anderson (ed.), *Encyclopedia of Hydrological Sciences*, John Wiley & Sons, New York, NY, USA.

Solomatine, D., Dibike, Y.B. and Kukuric, N. (1999). Automatic calibration of groundwater models using global optimization techniques. *Hydrological Science Journal*, 44(6), pp. 879-894.

Solomatine, D. and Dulal, K.N. (2003). Model trees as an alternative to neural networks in rainfall–runoff modelling. *Hydrological Sciences Journal*, 48(3), pp. 399-411.

Solomatine, D., Maskey, M. and Shrestha, D.L. (2006). Eager and lazy learning methods in the context of hydrologic forecasting, *Proc. of International Joint Conference on Neural Networks*, Vancouver, BC, Canada, IEEE.

Solomatine, D., Maskey, M. and Shrestha, D.L. (2008). Instance-based learning compared to other data-driven methods in hydrological forecasting. *Hydrological Processes*, 22(2), pp. 275-287.

Solomatine, D. and Shrestha, D.L. (2009). A novel method to estimate total model uncertainty using machine learning techniques. *Water Resources Research*, 45, W00B11, doi: 10.1029/2008WR006839.

Solomatine, D. and Siek, M.B. (2006). Modular learning models in forecasting natural phenomena. *Neural Networks*, 19(2), pp. 215-224.

Sorooshian, S. and Dracup, J.A. (1980). Stochastic parameter estimation procedures for hydrologic rainfall-runoff models: Correlated and heteroscedastic error cases. *Water Resources Research*, 16(2), pp. 430-442.

Spear, R.C. and Hornberger, G.M. (1980). Eutrophication in peel inlet-II. Identification of critical uncertainties via generalized sensitivity analysis. *Water Research*, 14(1), pp. 43-49.

Stravs, L. and Brilly, M. (2007). Development of a low-flow forecasting model using the M5 machine learning method. *Hydrological Sciences Journal*, 52(3), pp. 466-477.

Sudheer, K.P., Gosain, A.K. and Ramasastri, K.S. (2002). A data-driven algorithm for constructing artificial neural network rainfall-runoff models. *Hydrological Processes*, 16(6), pp. 1325-1330.

Sugawara, M. (1967). The flood forecasting by a series storage type model, *Proc. of International Symposium on floods and their computation*, Leningrad, USSR, IAHS Publication no. 85, pp. 1-6.

Sugawara, M. (1995). Tank Model. In: V.P. Singh (ed.), *Computer Models of Watershed Hydrology*, Water Resources Publication, Highlands Ranch, CO, USA, pp. 165-214.

Swart, R., Bernstein, L., Ha-Duong, M. and Petersen, A. (2009). Agreeing to disagree: uncertainty management in assessing climate change, impacts and responses by the IPCC. *Climatic Change*, 92(1), pp. 1-29.

Thiemann, M., Trosser, M., Gupta, H. and Sorooshian, S. (2001). Bayesian recursive parameter estimation for hydrologic models. *Water Resources Research*, 37(10), pp. 2521-2536.

Thyer, M., Kuczera, G. and Wang, Q.J. (2002). Quantifying parameter uncertainty in stochastic models using the Box-Cox transformation. *Journal of Hydrology*, 265, pp. 246-257.

Todini, E. (1988). Rainfall-runoff modeling ? past, present and future. *Journal of Hydrology*, 100(1-3), pp. 341-352.

Todini, E. (1996). The ARNO rainfall-runoff model. *Journal of Hydrology*, 175, pp. 339-382.

Todini, E. (2008). A model conditional processor to assess predictive uncertainty in flood forecasting. *Intl. J. River Basin Management*, 6(2), pp. 123-137.

Tung, Y.-K. (1996). Uncertainty and reliability analysis. In: L.W. Mays (ed.), *Water Resources Handbook*, McGraw-Hill, New York, NY, USA, pp. 7.1-7.65.

Uhlenbrook, S., Seibert, J., Leibundgut, C. and Rodhe, A. (1999). Prediction uncertainty of conceptual rainfall-runoff models caused by problems in identifying model parameters and structure. *Hydrological Sciences Journal*, 44(5), pp. 779–797.

Vapnik, V. (1995). *The nature of statistical learning theory*. Springer-Verlag, New York, NY, USA.

Vapnik, V. (1998). *Statistical learning theory*. John Wiley & Sons, New York, NY, USA.

Vrugt, J., ter Braak, C., Clark, M., Hyman, J. and Robinson, B. (2008a). Treatment of input uncertainty in hydrologic modeling: Doing hydrology backward with Markov chain Monte Carlo simulation. *Water Resources Research*, 44, W00B09, doi: 10.1029/2007WR006720.

Vrugt, J., ter Braak, C., Gupta, H. and Robinson, B. (2008b). Equifinality of formal (DREAM) and informal (GLUE) Bayesian approaches in hydrologic modeling? *Stochastic Environmental Research and Risk Assessment*, doi: 10.1007/s00477-008-0274-y.

Vrugt, J.A., Diks, C.G.H., Gupta, H.V., Bouten, W. and Verstraten, J.M. (2005). Improved treatment of uncertainty in hydrologic modeling: combining the strengths of global optimization and data assimilation. *Water Resources Research*, 41, W01017, doi: 10.1029/2004WR003059.

Vrugt, J.A., Gupta, H.V., Bouten, W. and Sorooshian, S. (2003). A shuffled complex evolution metropolis algorithm for optimization and uncertainty assessment of hydrologic model parameters. *Water Resources Research*, 39(8), 1201, doi: 10.1029/2002WR001642.

Wagener, T. and Gupta, H.V. (2005). Model identification for hydrological forecasting under uncertainty. *Stochastic Environmental Research and Risk Assessment*, 19(6), pp. 378-387.

Wagener, T., McIntyre, N., Lees, M.J., Wheater, H.S. and Gupta, H.V. (2003). Towards reduced uncertainty in conceptual rainfall-runoff modelling; Dynamic identifiability analysis. *Hydrological Processes*, 17, pp. 455-476.

Werbos, P. (1974). *Beyond regression: New tools for prediction and analysis in the behavioral sciences*. Ph.D. Thesis, Harvard University, MA, USA.

Whigham, P. and Crapper, P. (2001). Modelling rainfall-runoff using genetic programming. *Mathematical and Computer Modelling*, 33, pp. 707-721.

Witten, I.H. and Frank, E. (2000). *Data mining: Practical machine learning tools with Java implementations*, 239. Morgan Kaufmaan, San Francisco, CA, USA.

WMO (1975). *Intercomparison of conceptual hydrological models used in operational hydrological forecasting. Geneva.* Operational Hydrology Report No. 7, WMO No. 429, World Meteorological Organization, Geneva, Switzerland.

Wonnacott, T.H. and Wonnacott, R.J. (1990). *Introductory statistics*. John Wiley & Sons, New York, NY, USA.

Xie, X.L. and Beni, G. (1991). A validity measure for fuzzy clustering. *IEEE Transactions on Pattern Analysis and Machine Intelligence*, 13(8), pp. 841-847.

Xiong, L., O'Connor, K.M. and Guo, S. (2004). Comparison of three updating schemes using artificial neural network in flow forecasting. *Hydrology and Earth System Sciences*, 8(2), pp. 247-255.

Xiong, L. and O Connor, K. (2002). Comparison of four updating models for real-time river flow forecasting. *Hydrological Sciences Journal*, 47(4), pp. 621-640.

Xiong, L. and O'Connor, K.M. (2008). An empirical method to improve the prediction limits of the GLUE methodology in rainfall–runoff modeling. *Journal of Hydrology*, 349, pp. 115-124.

Xu, C. (2001). Statistical analysis of parameters and residuals of a conceptual water balance model–Methodology and case study. *Water Resources Management*, 15(2), pp. 75-92.

Yapo, P., Gupta, H. and Sorooshian, S. (1996). Automatic calibration of conceptual rainfall-runoff models: sensitivity to calibration data. *Journal of Hydrology*, 181(1-4), pp. 23-48.

Yapo, P.O., Gupta, H.V. and Sorooshian, S. (1998). Multi-objective global optimization for hydrologic models. *Journal of Hydrology*, 204(1-4), pp. 83-97.

Yen, B. and Ang, A. (1971). Risks analysis in design of hydraulic projects. In: C. Chiu (ed.), Stochastic Hydraulics, *Proc. of First International Symposium*, University of Pittsburgh, Pittsburgh, PA, USA, pp. 694-709.

Young, P. (1998). Data-based mechanistic modelling of environmental, ecological, economic and engineering systems. *Environmental Modelling and Software*, 13(2), pp. 105-122.

Young, P. and Ratto, M. (2009). A unified approach to environmental systems modeling. *Stochastic Environmental Research and Risk Assessment*, doi: 10.1007/s00477-008-0271-1.

Zadeh, L.A. (1965). Fuzzy sets. *Information and Control*, 8(3), pp. 338-353.

Zadeh, L.A. (1978). Fuzzy set as a basis for a theory of possibility. *Fuzzy Sets and Systems*, 1(1), pp. 3–28.

Zimmermann, H.-J. (1997). A fresh perspective on uncertainty modeling: Uncertainty vs. uncertainty modeling. In: B.M. Ayyub and M.M. Gupta (eds.), *Uncertainty analysis in engineering and sciences: Fuzzy logic, statistics, and neural network approach*, Springer, Heidelberg, Germany, pp. 353-364.

SAMENVATTING

Neerslag-afvoer modellen worden regelmatig gebruikt in de hydrologie voor een groot aantal toepassingen en spelen een belangrijke rol in de optimale planning en beheer van watervoorraden in stroomgebieden van rivieren. Een neerslag-afvoer model is per definitie een vereenvoudiging van de werkelijke complexe, niet-lineaire, tijd en ruimte gevarieerde, hydrologische processen. Dergelijke modellen bevatten parameters die niet vaak direct kunnen worden gemeten, maar die slechts kunnen worden geschat door calibratie met een historische reeks van gemeten outputgegevens. De gegevensreeksen van de systeeminput en de output bevatten vaak meetfouten. Daardoor zijn de voorspellingen die door zo'n model worden geproduceerd verre van perfect en inherent onzeker. Het is daarom essentieel dat de onzekerheid wordt erkend en dat er op de juiste manier rekening mee wordt gehouden. Zodra het bestaan van onzekerheid in een neerslag-afvoer model wordt erkend, moet deze worden geanalyseerd met behulp van de juiste onzekerheidsanalyse en voorspellingsprocedures die gericht zijn op het verkleinen van de effecten van onzekerheid. Er zijn een aantal van dergelijke procedures die actief worden gebruikt, maar onze analyse maakt het mogelijk om te concluderen dat deze vaak gebaseerd zijn op sterke veronderstellingen en dat ze bepaalde tekortkomingen hebben. Deze thesis is gewijd aan het ontwikkelen van nieuwe procedures voor onzekerheidsanalyse en -voorspelling en deze te testen op diverse afzonderlijke casussen.

In deze thesis onderzoeken we een aantal methoden die in de literatuur worden voorgesteld om zinvolle onzekerheidsgrenzen aan modelvoorspellingen toe te kennen. De meeste bestaande methoden bepalen de onzekerheid van de inputvariabelen door deze te propageren door het deterministische model en de output te analyseren en deze methoden vereisen daarom aannames over de inputverdeling en foutenstructuren. De meerderheid van de onzekerheidsmethoden past slechts één enkele onzekerheidsbron toe en negeert andere bronnen uitdrukkelijk. Geen enkele methode van schatting van de onzekerheid kan als perfect worden verondersteld in het juist weergeven van onzekerheid. Onze analyse leidt tot de conclusie dat de Machine Learning methoden die nauwkeurige modellen kunnen bouwen op basis van gegevens (in dit geval op gegevens van vorige modelfouten), uitstekend kunnen worden gebruikt als voorspellers van onzekerheid.

Machine Learning betreft het ontwerp en de ontwikkeling van algoritmen die computers toerusten om prestaties met de tijd te verbeteren op basis van gegevens. Het belangrijkste aandachtspunt van deze Machine Learning methoden is het automatisch opstellen van modellen met uit ervaring verkregen gegevens. De afgelopen 15 jaar zijn Machine Learning technieken gebruikt om Data-Driven (op gegevens gebaseerde) modellen van neerslag-afvoer te construeren. Deze technieken zijn tot op heden nog niet gebruikt om de onzekerheid van modelvoorspellingen expliciet te bepalen in de vorm van onzekerheidsgrenzen of kansverdelingen, in het bijzonder in de neerslag-afvoer modellering. Doel van dit onderzoek is de Machine Learning methoden te onderzoeken om onzekerheid in neerslag-afvoer modellering te analyseren, te modelleren en te voorspellen. We hebben twee methoden ontwikkeld, namelijk MLUE voor parametrische en UNEEC voor totale onzekerheidsanalyse van neerslag-afvoer modellen.

De Monte Carlo simulatie is een vaak gebruikte methode voor onzekerheidsanalyse in neerslag-afvoer modellering en staat de kwantificering toe van de model-outputonzekerheid als gevolg van onzekere modelparameters. De op Monte Carlo simulatie gebaseerde methoden voor onzekerheidsanalyse zijn flexibel, robuust, conceptueel eenvoudig en ongecompliceerd; methoden van dit type vereisen echter een groot aantal steekproeven (of model runs) en hun toepassing is soms beperkt tot eenvoudige modellen. Indien reken-intensieve modellen zouden worden gebruikt, zou de tijd en de middelen die deze methode vereist, buitensporig duur kunnen zijn. Een aantal methoden zijn ontwikkeld om de efficiency van de op de Monte Carlo gebaseerde onzekerheidsanalyse methoden te verbeteren en nog steeds vereisen deze methoden een aanzienlijk aantal runs om zowel in off-line als operationele modus een betrouwbare en zinvolle schatting van de onzekerheid te produceren. In deze thesis ontwikkelen we een methode om de parametrische onzekerheid van neerslag-afvoer modellen te voorspellen door Machine Learning methoden te ontwikkelen die de onzekerheidsresultaten van de Monte Carlo methode simuleren. De voorgestelde methode wordt aangeduid als MLUE (Machine Learning in parameter Uncertainty Estimation). De motivatie om de MLUE methode te ontwikkelen is om snelle parametrische onzekerheidsanalyse en voorspelling uit te kunnen voeren.

De MLUE methode is toegepast op een geaggregeerd conceptueel neerslag-afvoer model van het stroomgebied van de Brue (Groot-Brittanië). De Generalised Likelihood Uncertainty Estimation methode (GLUE, gegeneraliseerde waarschijnlijkheid onzekerheidsschatting) is gebruikt om de parameteronzekerheid van het model te analyseren. Wij passen de MLUE methode toe om de onzekerheidsresultaten te schatten die door de GLUE methode worden geproduceerd. We hebben laten zien hoe domeinkennis en de analysetechnieken worden gebruikt om de inputgegevens die in de MLUE methode worden gebruikt, te selecteren. Drie Machine Learning methoden, namelijk Artificial Neural Networks (kunstmatige neurale netwerken), Model Trees (model (beslis)bomen), en Locally Weighted Regression (lokaal gewogen regressie), zijn gebruikt om de onzekerheid van de modelvoorspellingen te voorspellen. In de MLUE methode zijn de prestaties van de Machine Learning methoden gemeten op grond van hun voorspellend vermogen (o.a., de correlatie coefficiënt, en de wortel van de gemiddelde kwadratische fout (RMSE)) en statistieken van de onzekerheid (o.a., de dekkingskans van voorspellingsgrenzen, en de gemiddelde voorspellingsgrenzen). Het is aangetoond dat de Machine Learning methoden de onzekerheidsresultaten met redelijke nauwkeurigheid kunnen voorspellen. Het grote voordeel van de MLUE methode is de efficiency waarmee de simulatieresultaten van het Monte Carlo systeem worden gereproduceerd; het kan hierdoor een effectief instrument zijn om in real time de onzekerheid van overstromingsvoorspellingen in te schatten.

Over het algemeen bestaat de analyse van onzekerheid uit het propageren van de onzekerheid van de input en van de parameters (welke aangegeven wordt met een verdelingsfunctie) door het model voldoende keren te runnen om zo de verdelingsfunctie van de model output te bepalen. Als onderdeel van dit onderzoek ontwikkelen we een nieuwe methodologie, die de onzekerheid van de optimale modeloutput analyseert, modelleert en voorspelt door historische modelfouten te analyseren. Deze methode noemen we UNcertainty Estimation based on Local Errors and Clustering (UNEEC, Onzekerheid Schatting gebaseerd op Locale Fouten en Groepering). De UNEEC methode bestaat uit drie belangrijke stappen: (i) het groeperen van de inputgegevens om de homogene gebieden van de inputruimte te identificeren, (ii) het schatten van de kansverdelingen van de modelfouten voor de geïdentificeerde gebieden, en (iii) het bouwen van de Machine Learning modellen van de kansverdelingen van de modelfouten. Fuzzy groepering is gebruikt om de inputgegevens te groeperen. Drie Machine Learning methoden zijn gebruikt om de onzekerheid van de modelvoorspellingen te voorspellen namelijk: Artificial Neural Networks (kunstmatige neurale netwerken), Model Trees (model (beslis)bomen), en Locally Weighted Regression (lokaal gewogen regressie)

De UNEEC methode wordt toegepast op neerslag-afvoer modellen van drie verschillende stroomgebieden: (i) Data-Driven modellen voor het stroomgebied van de Sieve, Italië; (ii) een geaggregeerd conceptueel neerslag-afvoer model voor het stroomgebied van de Brue, Groot-Brittanië; en (iii) een geaggregeerd conceptueel neerslag-afvoer model voor het stroomgebied van de Bagmati, Nepal. Het is aangetoond dat de met de UNEEC methode geschatte onzekerheidsgrenzen in overeenstemming zijn met de orde grootte van de modelfouten. De resultaten laten zien dat het percentage geobserveerde afvoerwaarden dat binnen de geschatte voorspellings- of onzekerheidsgrenzen valt, bijna gelijk is aan het gespecificeerde betrouwbaarheidsinterval dat is gebruikt om deze grenzen niet te breed te laten worden. We vergelijken de onzekerheid van de modelvoorspellingen met andere methoden om onzekerheid te schatten, namelijk GLUE en twee statistische methoden, meta-Gaussiaans en kwantitatieve regressie. De vergelijkingsresultaten tonen aan dat de UNEEC methode consistent smallere onzekerheidsgrenzen genereert.

In dit onderzoek hebben we ook de Multi-Objective (voor meerdere doelfuncties) optimaliseringsroutine NSGA-II toegepast om het HBV neerslag-afvoer model te calibreren voor het stroomgebied van de Bagmati. Vier doelfuncties die verschillende aspecten van de afvoer hydrograaf benadrukken zijn de volumefout, de wortel van de gemiddelde kwadratische fout (RMSE), de RMSE van lage afvoeren, en de RMSE van piekafvoeren. We passen Pareto-voorkeur sortering toe om een klein aantal oplossingen te selecteren uit de vier-dimensionale Pareto-optimale oplossingen. De geselecteerde Pareto-optimale reeks wordt vergeleken met de optimale waarden van de parameters voor elk van de vier afzonderlijke doelfuncties. De resultaten laten zien dat de geselecteerde Pareto-optimale reeks een goed compromis is tussen alle vier de doelfuncties, en binnen de beste en slechtste oplossingen van elke afzonderlijke doelfunctie-optimalisatie valt. Multi-Objective calibratie staat ook de kwantificering van de onzekerheid toe in de vorm van de verschillende modelsimulaties die overeenkomen met de Pareto-optimale parameter sets; en het verstrekt extra informatie die nodig is voor besluitvorming op grond van de risicobenadering.

Dit onderzoek heeft aangetoond dat Machine Learning methoden redelijk nauwkeurige en efficiënte modellen kunnen bouwen voor het voorspellen van onzekerheid, in het bijzonder in neerslag-afvoer modellering. Wij komen tot de conclusie dat de Machine Learning technieken een waardevol instrument voor onzekerheidsanalyse van diverse voorspellingsmodellen kunnen zijn. Het onderzoek geeft ook aanbevelingen voor toekomstig onderzoek en ontwikkeling voor onzekerheidsanalyse op het gebied van neerslag-afvoer modellering.

Durga Lal Shrestha

Delft, The Netherlands

ACKNOWLEDGEMENT

I would like to express my sincere thanks to my supervisors Prof. Dimitri P. Solomatine and Prof. Roland Price for their continuous guidance and support with great patience in making this research possible. Prof. Dimitri, I am greatly indebted to you for your effort to stimulate and formulate this research within EU project. Your creative and dynamic inspiration encouraged me throughout this research period. You not only provided me scientific support, but also social and personal support. Prof. Price, you always remained helpful and available when I needed you. Several invaluable and inspiring discussions with you have helped to improve this research. I have learnt and benefited a lot from you two. I enjoyed very much working with both of you.

This research project was financed by the European Community's Sixth Framework Program through the grant to the budget of the Integrated Project FLOODsite, contract GOCE-CT-2004-505420. I enjoyed very much working with the FLOODsite partner. During the project period I had opportunities to discuss my research with Prof. J. Hall and H. Harvey, Newcastle University, UK, Prof. M. Borga, University of Padova, Italy, P. van Gelder, TU Delft, Netherlands, Prof E. Todini, University of Bologna, Italy. I am very thankful to all of them.

It has been great opportunity to conduct my research with the Department of Hydroinformatics and Knowledge Management at UNESCO-IHE. All the staff members are very friendly and supportive. I would like to thank all of them for the informal and fruitful discussions I had with them during the research period. I am grateful to Ir. Jan Luijendijk, head of the department, for supporting and providing me the nice working space and environment in the department premises. I am also thankful to all the staff of UNESCO-IHE; they are very cooperative and provided me help when I needed. I am also grateful to Jolanda Boots for assisting in administrative and practical matter of this research, Laura Kwak for co-ordinating at the last stage of the research period, Peter Stroo for managing to print out this thesis. I would like to thank Vera Schouten and Schalk Jan van Andel for helping to prepare the Dutch translation of the summary of this thesis.

I would like to thank to Department of Hydrology and Meteorology, Kathmandu, Nepal for providing opportunity first to do M.Sc. in Hydroinformatics at UNESCO-IHE and then extending my leave to conduct this research. In particular, I am grateful to former Director General Aadarsh P. Pokharel and Deputy Director General Keshav Sharma for their support and encouragement. The Natural Environment Research Council, Centre for Ecology and Hydrology, Wallingford, [Moore, R.; Pedder, M.; Cluckie, I.] is thanked for making the HYREX data set available via the British Atmospheric Data Centre Web site at http://badc.nerc.ac.uk/data/hyrex/.

I would like to thank the members of the PhD awarding committee, in particular, Prof. S. Uhlenbrook (UNESCO-IHE), Prof. A. Montanari (University of Bologna), Prof. J. Hall (Newcastle University), Prof. N. van de Giesen (TU Delft) for their invaluable comments and suggestions.

My genuine thanks go to the PhD colleagues at UNESCO-IHE with whom I shared my experiences, sufferings and joys. In particular, I always cherish the memory of sharing the office together and research ideas with PhD colleague Gerald Corzo. My sincere appreciation is also due to all of my M.Sc. students who helped me directly or indirectly to complete this research. I would like to thank to the Nepalese friends and seniors in Delft and in the Netherlands for reminding our culture and festivals by organizing social events and gatherings. I am also grateful to Dutch friend Marie-José van der Sandt for helping the Dutch translation of the summary of this thesis. There are many others to name and thank, but may be it is wiser not to forget them.

Finally, I am very thankful to my family members, relatives and friends in Nepal, in particular my mother for constant deep love, brothers and sisters for encouragement, parents-in-law for continuous support. Above all, my special and warm thanks go to my wife Srijana and son Sujan for their patience and love. Srijana, you have provided me unconditional support to complete this research smoothly. You suffered more than me. This journey would not be possible without your encouragement, inspiration and scarification.

ABOUT THE AUTHOR

Durga Lal Shrestha graduated in civil engineering with distinction from Tribhuvan University, Nepal in July 1996. He worked as consultant engineer in private engineering consultancy from April 1996. In November 1996 he joined Department of Hydrology and Meteorology under Ministry of Science and Technology at Kathmandu, Nepal as hydrologist engineer. During this time his major responsibilities were design, monitoring and networking of hydrological stations; analysis and processing of hydrological data; design, development and management of hydrological database. In 1998 he was also involved as project engineer for Nepal Irrigation Sector Project. In October 2000 he came to IHE (now UNESCO-IHE), delft to follow the Master's program in Hydroinformatics. He received his M.Sc. degree in Hydroinformatics with distinction in April 2002. From May 2002 to February 2004 he worked as researcher in Delft Cluster project in Department of Hydroinformatics and Knowledge Management at UNESCO-IHE. During this period his research work was mainly related to application of data-driven modelling in water systems. Since March 2004 he has been at UNESCO-IHE as a PhD student; involved in EU Sixth framework project FLOODsite (March 2004-Februray 2009); and assisted in various ways in the Hydroinformatics Master's Programme. From June, 2009 he continued to work as a Postdoctoral researcher in the Department of Hydroinformatics and Knowledge Management.

During the research period he published more than 20 technical papers in international journal and conferences. He is a member of the International Association of Hydrological Sciences (IAHS), International Association of Hydraulic Research (IAHR), American Society of Civil Engineers (ASCE), IEEE, Computational Intelligence Society, European Geosciences Union, the Nepal Engineer's Association (NEA) and the Nepal Engineer Council (NEC).

Publications List (not all of them are related to this thesis)

Peer Reviewed International Journals

Shrestha, D.L., Kayastha, N. and Solomatine, D. (2009). A novel approach to parameter uncertainty analysis of hydrological models using neural networks. *Hydrology and Earth System Science,* 13, pp. 1235-1248.

Solomatine, D. and **Shrestha, D.L.** (2009). A novel method to estimate total model uncertainty using machine learning techniques. *Water Resources Research*, 45, W00B11, doi:10.1029/2008WR006839.

Solomatine, D., Maskey, M. and **Shrestha, D.L.** (2008). Instance-based learning compared to other data-driven methods in hydrological forecasting. *Hydrological Processes*, 22(2), pp. 275-287.

Shrestha, D.L. and Solomatine, D. (2008). Data-driven approaches for estimating uncertainty in rainfall-runoff modelling. *Intl. J. River Basin Management*, 6(2), pp. 109-122.

Shrestha, D.L. and Solomatine, D. (2006). Experiments with AdaBoost.RT, an improved boosting scheme for regression. *Neural Computation*, 18(7), pp. 1678-1710.

Shrestha, D.L. and Solomatine, D. (2006). Machine learning approaches for estimation of prediction interval for the model output. *Neural Networks*, 19(2), pp. 225-235.

Conference Proceedings

Shrestha, D.L., Kayastha, N. and Solomatine, D. (2009). Parametric uncertainty estimation of a hydrological model using piece wise linear regression surrogates, *Proc. of the 33rd IAHR Congress*, IAHR, Vancouver, Canada, August 9-14, 2009.

Shrestha, D.L., Kayastha, N. and Solomatine, D. (2009). ANNs and other machine learning techniques in modelling models' uncertainty, *Proc of 19th International Conference on Artificial Neural Networks*, September 14-17, Limassol, Cyprus.

Shrestha, D.L., Kayastha, N. and Solomatine, D. (2009). Encapsulation of Monte-Carlo uncertainty analysis results in a predictive machine learning model, *Proc. of the 8th International Conference on Hydroinformatics*, Chile.

Adel, A., Shrestha, D.L., van Griensven, A., and Solomatine, D. (2009). Comparison of calibration and uncertainty analysis methods: case study of Nzoia River SWAT model, *Proc. of the 8th International Conference on Hydroinformatics*, Chile.

Shrestha, D.L. and Solomatine, D. (2008). Uncertainty modelling in flood forecasting: application of data driven models, *Proc of the Floodrisk 2008*, Oxford, UK.

Shrestha, D.L. and Solomatine, D. (2008). Comparing machine learning methods in estimation of model uncertainty, *Proc. of the International Joint Conference on Neural Networks*, IEEE, Hong Kong, China.

Shrestha, D.L. and Solomatine, D. (2007). Predicting hydrological models uncertainty: use of machine learning, *Proc. of the IAHR Congress*, IAHR, Venice, Italy.

Shrestha, D.L. and Solomatine, D. (2006). A Novel method to estimate the model uncertainty based on the model errors. In: A. Voinov, A.J. Jakeman and A.E. Rizzoli (eds.), Proc. of the *iEMSs Third Biennial Meeting: "Summit on Environmental Modelling and Software"*, International Environmental Modelling and Software Society, Burlington, USA.

Solomatine, D., Maskey, M. and Shrestha, D.L. (2006). Eager and lazy learning methods in the context of hydrologic forecasting, *Proc. of the International Joint Conference on Neural Networks*, IEEE, Vancouver, BC, Canada.

Shrestha, D.L., Rodriguez, J., Price, R.K. and Solomatine, D. (2006). Assessing model prediction limits using fuzzy clustering and machine learning, *Proc. of the 7th International Conference on Hydroinformatics*, Research Publishing, Nice.

Changjun, C., Shrestha, D.L., Corzo, G. and Solomatine, D. (2006). Comparison of methods for uncertainty analysis of hydrologic models, *Proc. of the 7th International Conference on Hydroinformatics, Research Publishing*, Nice, pp. 1309-1316.

Shrestha, D.L. and Solomatine, D. (2005). Estimation of prediction intervals for the model outputs using machine learning, *Proc. of the International Joint Conference on Neural Networks*, IEEE, Montreal, Canada, pp. 2700-2705.

Shrestha, **D.L.** and Solomatine, D. (2005). Quantifying uncertainty of flood forecasting using data driven modeling, *Proc. of the 29th IAHR Congress*, IAHR, Seoul, South Korea, pp. 715-726.

Solomatine, D. and **Shrestha, D.L.** (2004). AdaBoost.RT: a boosting algorithm for regression problems, *Proc. of the International Joint Conference on Neural Networks*, IEEE, Budapest, Hungary, pp. 1163-1168.

Bhattacharya, B., **Shrestha, D.L.** and Solomatine, D. (2003). Neural networks in reconstructing missing wave data in sedimentation modelling, *Proc. of the XXXth IAHR Congress*, THEME D: Maritime Hydraulics, Thessaloniki, Greece, pp. 209-216.

Presentation and Research Abstracts

Shrestha, D.L., Kayastha, N. and Solomatine, D. (2009). A novel approach to parameter uncertainty analysis of hydrological models: Application of machine learning techniques, *Proc of the Geophysical Research Abstracts*, European Geosciences Union.

Shrestha, D.L., Kayastha, N. and Solomatine, D. (2009). A novel approach to Monte Carlo-based uncertainty analysis of hydrological models using artificial neural networks, *Proc of the Geophysical Research Abstracts*, European Geosciences Union.

Shrestha, D.L. and Solomatine, D. (2008). Novel method to estimate predictive uncertainty of flow forecasting, *Proc of the Geophysical Research Abstracts*, European Geosciences Union

Shrestha, D.L. and Solomatine, D. (2007). Comparing machine learning approaches in estimating model uncertainty of hydrological conceptual models, *Proc. of the Geophysical Research Abstracts*, European Geosciences Union.

Shrestha, D.L. and Solomatine, D. (2007). Neural networks and clustering in estimation of the total model uncertainty of hydrologic models, *Proc. of the Geophysical Research Abstracts*, European Geosciences Union.

Shrestha, D.L. and Solomatine, D. (2006). Fuzzy clustering and neural networks in localized estimation of the total model uncertainty, *Proc. of the Geophysical Research Abstracts*, European Geosciences Union.

Solomatine, D., **Shrestha, D.L.** and Chen, C. (2006). Estimating parameter uncertainty of hydrological models by Metropolis-Hastings, SCEM-UA, and Adaptive cluster covering (ACCO) algorithms, *Proc. of the Geophysical Research Abstracts*, European Geosciences Union.

Solomatine, D., Bhattacharya, B. and **Shrestha, D.L.** (2005). Data-driven modelling vs. machine learning in flood forecasting, *Proc. of the Geophysical Research Abstracts*, European Geosciences Union.

For Product Safety Concerns and Information please contact our EU representative GPSR@taylorandfrancis.com Taylor & Francis Verlag GmbH, Kaufingerstraße 24, 80331 München, Germany

T - #0139 - 160425 - C56 - 244/170/15 - PB - 9780415565981 - Gloss Lamination